Science In Civil Society

Science In Civil Society

John Ziman

imprint-academic.com

Published in the UK by
Imprint Academic, PO Box 200, Exeter EX5 5YX, UK

Published in the USA by
Imprint Academic, Philosophy Documentation Center
PO Box 7147, Charlottesville, VA 22906-7147, USA

ISBN-13: 978 184540 0828

A CIP catalogue record for this book is available from the
British Library and US Library of Congress

Contents

Joan Solomon

Preface

It is obviously and gloriously right that John Ziman, the author of this book, should be found presenting to the reader such a very vigorous and wide-ranging work. His arguments go a long way towards giving shape to a life-long sequence of thought. By the time that he started on writing this present book he had already written about the nature of science from a set of quite different perspectives, and indeed he had done it many times over, always in ways that were different and very far from *dull*. Nothing that he wrote could ever justify such a description by the use of that grim monosyllable. In this, his final book, he appealed right across the sequences of ideas that he had worked on throughout his life. So, in 2001, when he felt called upon to explain the social nature of his beloved science, he did so through the medium of this last book, working on some of the later points of it during the time of his final illness in hospital. Here, as we shall see, he was busy describing yet another set of ideas about science, embedded in yet further ways of looking at how science could be seen to be affecting society, and inversely, of how sociology could be seen to be affecting science.

Is science really 'social'?

In his first book about the operations of science in the social realm, *Public Knowledge*, Ziman — then a young and brilliant neophyte physicist — described how one could impatiently set aside the need to spend five years and a substantial amount of money in the construction of a piece of experimental or literary work in order to obtain a doctorate. Such diversions could easily take the place of just reading books and articles from science journals. He had already been regularly pulling down books of a wide variety from the library shelves. In the concluding sentences of this early book he admitted to the overwhelming power of a network of scientists working together.

> There may perhaps be better social means than that of the Stock Exchange for maintaining the economic health of the nation; ... (but) ... I doubt if there is any better instrument for achieving a reliable knowledge of the world than *a freely cooperating community of scientists* (Ziman 1968).

That extract came neither from a book of scientific or economic data, nor from John Ziman's own mathematical calculations, such as those about the nature of ferromagnetism and other similar scientific systems that he rather cheekily called *Models of Disorder* (1979) in order to point out the contrast. Such knowledge systems, derived from accounts of straightforward ways of proceeding, and through which scientists could recognise the much-discussed *social nature of knowledge* or indeed of the newly emerging *cult of 'postmodernism'*. Naturally enough, this was not much taken up by the senior members of the scientific establishment at first, but Ziman wrote that it was particularly important to realise that science, far from being carved deeply into immutable granite inscriptions, was in the business of acquiring essentially fluid and protean ways of being and thinking. So, if we accept that science is concerned with everyday life, then the epistemology of science,was to be the *Reliable Knowledge* that — as we will see — Ziman was to publish in 1978. That was about everyday science and will inevitably come in many different colours and contrasting modes of talk in the course of their discussions. It was a difficult a time in the philosophy of science as it always would be in the exploration of yet more everyday ways of thinking about it.

None of these problems perturbed the young John Ziman. He was entirely ready to accept and to peer more closely into any of the new possibilities of the epistemology of science. Indeed he did this with much vigour, like a gardener working in his plot of land (where Ziman himself was often to be seen digging and planting in and out of the seasons of the year). He soon found that the growth of plants might become not only more vigorous but also more varied in ways that were quite different from the usual ones. Of course there had always been new plots of scientific territory to tend, but none of the new scientific specialities could be expanded by so large a factor at the expense of all the others. It was not enough, Ziman wrote, to obtain a whole lot of new answers to old questions. In the fields of science it might be more a question as to what crop was needed to be cultivated at the present time. That should, at best, show up future possibilities for progress into new fields. The sciences of our society, together with the results that can almost always be obtained from them, may still be quite beyond our present guesses however many conflicting results may be obtained as outcomes while the scientists are working at it.

> It is particularly important to realise that science is fallible. Men are often fallible, and it is no more impossible, in principle, for a whole group to be deceived than it is for one frail mortal to make a mistake. The only advantage that science has is that it is like the old Polish parliament: everyone has a veto and is trained to exercise it (Ziman 1981, p. 32).

In ways that he was to explore much more thoroughly in his later books, Ziman was already beginning to face up to the question of how to use the occasionally awkward characteristics of any particular method through which scientific knowledge might possibly be expanded and fashioned anew.

Shedding light on *Public Knowledge*

I have to admit that my own introduction to Ziman's ever increasing number of books about the nature of science presented a more difficult reading task than the earlier ones had done. But then it was also more rewarding for the reader. At first *Public Knowledge* does not seem to be about science at all, but persevering just a minimum amount reveals a subtitle which is much more helpful: *An essay concerning the social dimension of science.* That word 'social' again! The year was now 1968 and John had been awarded a fellowship of the Royal Society—a social reward if ever there was one—which he accepted with a pride in the dizzy heights of this commendation that he continued to relish throughout his life. Nevertheless a persevering reader might already have tracked down, from the title page or elsewhere, that this early interest in the social features of science was still guiding his work. There would, from now on, be a social dimension to almost every line of this and most of his later books. That in itself called for considerable commendation of the work of such a young scientist. *But what were these new 'social dimensions'?*

It was still too early in John's self-appointed mission to begin understanding the complete nature of science, but this small book—*Public Knowledge*—did, at the very least, make a splendid beginning. Ziman had, quite simply, decided to take apart, very carefully, the nature and emergence of scientific knowledge with all its social features. That involved distinguishing the scientific method from the scientific argument, a scientific life from a scientific education, an individual scientific recipe from a scientific communication, and the scientific institutions from the scientific authorities. In retrospect we can see that the recipe had begun to work and, as he seems to have guessed, that he was on the right track.

> The real issue is that the distinction between formal and informal scientific communications should not be blurred. The official scientific paper in a reputable journal is not an advertisement or a news item; it is a contribution to *the consensus of public knowledge* ... A major achievement of our civilisation is the creation of this form of communication, however clumsy and barbaric it may seem to those whose concerns are with poetry and feeling. The individual primary (scientific) paper is not the final form of the consensus but it is a brick from which the whole edifice is to be built (p. 109).

That is not to say that the method always worked, nor even that it would always be possible to distinguish the scientific from the non-scientific, but Ziman was making progress frequently enough for the method to lodge excitingly in his mind, until about thirty two years later. That was when his most important, and quite encyclopaedic book—*Real Science* (2000)—was published and its readers, including myself, could begin to understand that we were several rungs higher up the ladder than they, or others, had ever been before. It is intriguing to read how, because of his love of fine literature and poetry, John—mathematician, scientist, and avid reader in many regions—felt the need to add another sentence about literature to the excerpt above: 'Perhaps also I may have shown that Science itself is by no means as inhuman as it is sometimes painted' (p. 142).

Approaching education in science

Of course education is central to the task of building up, in each generation, a reflection of the community's picture of science, and then to contrive its use in ways that may be enhanced by the work of a coming generation of research students. *Public Knowledge* had written under its main title '*Concerning the social dimension of science*' and he argued over it in many parts of the text. The problem was not that the work was too difficult to understand at the high academic level that Ziman chose to use; rather that it ran at the higher level of engagement in academic lecturing where no one checked carefully enough on its effectiveness as education. Many of the lecturers, even in the most prestigious of universities, had little ability in teaching well (or on some occasions, even just being audible to students) and in a manner tailored to fit into first class education. John Ziman was well aware of this problem. He was himself a fine lecturer—but that did not make it any easier to pass on his skills and understanding to another lecturer. It was unfortunate that, as he wrote in *Public Knowledge*, the very topics which he expected to introduce into the undergraduate physics course, were the most difficult to explain. He took a hard line on this.

> My own feeling that is that the History and Philosophy of Science have already become so academicized as to be meaningless to the average young scientist (p. 66).

So Ziman changed the object of his search and looked instead for a communal system in which each student's independent hunt for something like to the ancient philosopher's stone, might become a set of systems from which all must jump together when either entering or leaving the communal train. There could be no great enthusiasm for that as a model for teaching and learning. While the students' minds were so firmly fixed on the looming examinations, and the likelihood of getting a first class degree, they often continued to have little time for anything else.

He tried to find a more successful way into education, and in that he succeeded by favouring the use of 'the context of imagination' rather than that of 'creativity'. He was especially successful in the generation of new patterns of discovery out combinations of ideas. Indeed this book itself had been dedicated to his philosopher and friend Norwood Russell Hanson, who had died recently before this book was completed, after publishing a small but much admired book entitled *Patterns of Discovery*. Ziman explained his use of the word 'imagination' as a foil to the more slippery concept of creativity in education (see M. Boden 1994). The term 'scientific imagination' is often used for the power that enables a student of science to muse, for example, about a sample of an ideal gas which was being warmed up, in terms of its pressure and 'mean free path' within the gaseous turmoil of collisions. This was an alternative to finding the root mean square velocity of its 'ideal but invisible' molecules. The whole idea appealed greatly to Ziman who compared this way of thinking with the cultivation of poetic or artistic genius. As a lifelong teacher of science I too could relish the three qualities that John Ziman used in his final summing-up of the directions he saw as leading to a creative education in science. He wrote about their great value, and then wisely left the crafting of practical details to the teacher.

> Learning, imagination and critical (common) sense ... are the three qualities which the scientific mind must possess in abundance (Ziman 1968, p. 80).

Technology and the *Force of Knowledge*

In his earlier book on *Public Knowledge,* John Ziman had written briefly but valuably about the distinction between Science and Technology. An expansion of this was to be the foundation of his next and much longer book — *The Force of Knowledge* (1976). Once again the title contains an additional phrase that is shedding important extra light on his purposes. What did he mean by 'The Scientific Dimension of Society'? Ziman wrote that the word 'science' was to be taken to mean 'The Art of *Knowing*', and elsewhere that 'technology' could be translated as 'The Art of *Knowing How*'. The reader may need to allow these two rather strange phrases freedom to illuminate much of the work as it is being read. As far as music was concerned Ziman was no artist but he enjoyed listening to the works of Mozart. However he was *very* susceptible to visual signals from both human works of art, and those of nature. Indeed I once found him standing stock still and silent, some twenty yards behind me, on the cliff top, where he seemed quite immobilised by the beauty of rare wild flowers against the background of fierce breaking waves, all in South Africa's Cape Town Silver Mine. He always seemed able to a combine an appreciation of scientific knowledge with what he saw of the beauty of natural surroundings. The invention of new machinery, intricate mathematics, the appreciation of scientific discovery, as well as the phenomena of nature, could all stimulate deep and delighted reactions.

Reading more deeply into *The Force of Knowledge* shows clearly what is to be gained by combining art with architecture, engineering with literature, and often he presented them by a range of simple line-drawn illustrations, cartoons and photographs. It was an exciting approach to the multitude of human ways for acquiring scientific knowledge, and was also full of appropriate contemporary illustrations down to the last detail. Much of the practical sociology that can be found at work within science has always been represented by contemporary art, and the emergence of new cultures could usually be compared with pictures of much older ones. In the present context all of this can be digested

and explained to illustrate Ziman's enthusiasm for technological discovery, using ancient woodcuts, or more
contemporary political cartoons, and even by some quite
new informational methods. Even nets and sketches, and
Feynman diagrams are put to explanatory use in his postgraduate lectures, producing a more theoretical book, *Elements of
Advanced Quantum Theory* (1969). All of these produce a concerted understanding wherever new sciences were being helped
along by the use of the internet. The double spread of illustrations
from a geological map *on the left* (1976, p. 54) that had been
painted by the impoverished William Smith—the first canal
builder, who had died in a debtor's prison—to, *on the right*, a
crowd of ultra well- dressed members of the Royal Society in a
contemporary *Conversazione*, all pictured examining the most
recent state-of-the art microscopes. Ziman sometimes called this
phenomenon by which scientific knowledge becomes ever
more theoretical as the 'academicization' of science. He was to
write much more on this theme in the very last of his books,
posthumously published here as *Science and Civil Society*.

At approximately the same time as the newly published
Force of Knowledge appeared on the scene, Ziman published
two more books, both of which fitted well into his varied
current interests. *Reliable Knowledge* came out in 1978 and
provided a term which will always be associated Ziman's
attitude towards what others might have called something
like 'Can Science be trusted?' But that was not in the spirit of
his more emphatic belief and trust in proven academic
excellence. The term 'reliable knowledge' has caught on
surprisingly well with the general public, where it may be
providing some comforting messages about the safety of
our general public.

The Force of Knowledge was a very successful and marketable book. As Ziman himself acknowledged privately, it
almost ran the risk, in parts, of becoming a slightly superior
'coffee-table book'! Its pages are full of recommendations
for further reading. No doubt this was at least partially due
to the large number of illustrations that had been so carefully chosen to illustrate the nature of science in different
historical contexts. Every chapter not only had a clear title,

it also had acquired a printed section at the back of the book which suggested suitable 'Questions and Answers' for the use of students and teachers. This multiple method of supplying references to guide students and their teachers shows Ziman's concern with the improvement of science education and runs valuably throughout the whole of this book.

Finally I pick out for special mention the huge increase in the number of specialised academic journals (p. 107) from the 1930s until present times. That happened before the Internet took over the bulk of this ever increasing publishing job, and so reduced the weight of paper required. Warfare and the granting of money for research into the use and manufacture of weapons of offence also takes up several chapters, again with a suitable selection of illustrations. Some readers may ask, 'Where were the much quoted sociological points of view?' As we have seen, John Ziman was always on the look-out for links between science and sociology. Was it now possible to show how these two disciplines may be seen together?

The sociologies of scientific knowledge

Bringing these two areas of study together — the sociological and the scientific — was still to be tackled. If people who had been trained in any specific branch of scientific knowledge then continued to work together, there would always be a strong possibility that they would come to share some of the same habits of thinking and learning. There would be nothing strange in such an outcome. Ziman claimed that it was at its most valuable in traditional internalist accounts of scientific discovery. In the preface to the *Force of Knowledge* Ziman had claimed that he had written mostly for students who had a background in physics, philosophy or sociology. The starting point for his argument about the weak and strong sociologies of science, was to see them as populating internal worlds where learning about the history and discoveries of science may take the working scientist through a range of familiar routes and back into the narrower *academic* world of science.

However another kind of sociology was also increasingly at work in industry and in medicine, where the power to change how science was producing new outcomes, we might see as new pressures upon the external sociology of science. The inventive powers of scientists working with the new forms of science and technology could now be expanded outwards into what Ziman was to study separately, as the *external sociology of science*. This included the 'collectivization' of science (1997) into new systems of research and development.

For any book to study the new sociologies of science, and to do the job carefully and critically enough to guide readers through the big changes that they would be bound to meet, would be a hard task. Ziman had begun the work started as *The Force of Knowledge* where he showed that the traditional mode of scientific research was not 'just a method' nor was it 'all about public esteem' brought about, perhaps by peer review. But it was still concerned with such matters, not because of their fundamental meaning, which so very few of the public understood, but because the old 'academic' ways of driving modern science were pumping ever more uncertain expectations into new fields of the sociology of science. Many of these were concerned with increasing industrial wealth, which cut across the disinterested stance of Merton's (1973) sociology of science.

While John Ziman's previous books had all concentrated on the familiar historical and philosophical trends in the sociology of science, his next one — *Real Science* (2000) — focussed instead on a view of the emerging network of knowledge production which would grow into a powerful mixture of both the *internal and external sociologies* of science. Ziman saw at once that such a book would not only be difficult to write but also to read. It was his intention to use a language that was direct and simple, what he called 'lay language', to describe the different 'schools' of sociology, as might have been used in mediaeval times. Only then would his book show, in his own words, the sense of 'What Science is, and what it means'. Here another point needs to be made about the nature of *Real Science* before we jump in for a cold shower of reality. Ziman knew that Science was under siege and, worse still, that the tendency to attack one or more of its science sociologies would come from the very people who had previously been so ready to jump to their defence.

This is not the place to write much more about these emerging sociologies of science. Ziman's book about this lies spread open in front of me. All I would want to add is that terms like 'culture' often take the place of 'science' whether or not the wicked addition of the preface 'meta' has been included. It has been common to refer to the whole of the science culture as a 'knowledge community'. The scientists themselves always tend to treat science knowledge as an ethically self-organising system where the mere suggestion that they might need independent help raises nothing but anger. It was the American sociologist of science Robert Merton (1973) who first laid out the qualities that working scientists would need — as if they had not already got them — in order to be 'good scientists'. All of these are discussed in *Real Science* (2000).

And then, not surprisingly, we had the 'Science Wars' between those for whom science was not only how scientists thought — better than anyone else — but also how far they resembled those who seemed, at first sight, new and immaculate knights of the Round Table.

Joan Solomon
University of Plymouth

References

Boden, M.A. (ed. 1994). *Dimensions of Creativity*. MIT Press.

Hanson, N.R. (1958). *Patterns of Discovery*. CUP

Merton, R.K. (1942 [1973]). The Normative Structure of Science. *The Sociology of Science*. ed. N.W. Storer. U of Chicago Press.

Ziman, J. (1968). *Public Knowledge: An essay concerning the social dimension of science*. CUP.

Ziman, J. (1969). *Elements of Advanced Quantum Theory*. CUP

Ziman, J. (1976). *The Force of Knowledge: The scientific dimension of society*. CUP.

Ziman, J. (1978). *Reliable Knowledge*. CUP.

Ziman, J. (1979). *Models of Disorder* . CUP.

Ziman, J. (1981). *Puzzles, Problems and Enigmas: Occasional pieces on the human aspects of science*. CUP.

Ziman, J. (1997). *Of One Mind: The Collectivization of Science*. Springer.

Ziman, J. (2000). *Real Science: What it is and what it means*. CUP.

ONE

What Is Science For?

What do we mean by science?

These days, science is everywhere. It pervades our whole society. People meet it in a multitude of forms, around every corner, in every sphere of life. Sometimes it is just a clutter of commonplace frivolities, like new fashion fabrics. Sometimes it miraculously preserves our life, like penicillin. Sometimes, like climate change, it looms over us as a portent of doom: sometimes it promises a way of escape from such a fate. Sometimes, like a nuclear warhead, it enshrouds us in political terror: sometimes, like a verification technology, it offers an antidote to such evils. Sometimes it presents itself as cool logic and sometimes as a mystical incantation. Sometimes, in its passion for counting and classifying, it seems utterly dreary: sometimes it beguiles us with curious and wonderful ideas.

Such episodes are entirely familiar, and endlessly diverse. Their only common feature is that they are encounters with 'science'. *What does that mean?* Every time I start another book on the subject, I feel bound to pose that question. And then the whole work is in danger of never getting much further than sketching out another tempting path towards what may seem a plausible answer. And my shelves are overflowing with books pointing in other directions through the fog.

The nature of science seems to depend on what sort of thing one eventually expects to say about it. There are indeed many different ways of approaching it. Mostly, as a pensioner from a career in very basic research, I have been

looking from the inside.[1] We were in the business of producing 'scientific knowledge'. But the results of deep study
and high theory often looked quite incredible. Frequently
they didn't even justify themselves by how well they
worked in practice. So the challenge was to explain to
non-scientists what made some of our extraordinary discoveries seem so worthy of belief.

But starting from the centre like that, I never reached the
places where science meets with ordinary people. So this
time, I am coming at it from the outside. That requires a
much wider view, zooming out to cover the larger landscape of human life. What would science look like if one
were parachuted down into it from Mars (where presumably they are long past such prehistoric pursuits)?

No, that is just a sociological fantasy. People don't
encounter 'science' as a whole. They meet different aspects
of it, and in many different roles — as customers, patients,
clients, combatants, officials, journalists, victims, employers, etc. They come to it in various moods — practical, grateful, fearful, respectful, suspicious, accepting, rejecting, and
so on. It is shrouded in myths, camouflaged by political ideologies and blanketed in official secrets. Scientists themselves tell different stories according to the particular
segment of the public with which they have to deal.

So that is why generalised discourse about 'public attitudes towards science', or 'the place of science in modern
society', is so unrewarding. Like 'money' or 'law', 'science'
is integrated seamlessly into our life style. What is more, it is
a fuzzy concept, impossible to define precisely without
begging whatever question is being asked about it. Indeed,
its rhetorical uses are so notorious that worldly wisdom
advises one to suspect anyone who claims it for their own.
Some of us have learnt to conjugate it sceptically: 'I am
scientific; you have some interesting ideas; she is trying to
pull the wool over our eyes.' But that would be hopelessly

---

[1]   Ziman, J. M. 1978 *Reliable Knowledge: An Exploration of the Grounds for
       Belief in Science.* Cambridge: Cambridge UP, Ziman, J. M. 2000 *Real
       Science: What it is and what it means.* Cambridge: Cambridge UP.

counter-productive for any search which focused upon science within our broad civil society.

Generally speaking however, one of the most prominent features of our science is that it always fits into an *organised* body of knowledge. It is a collection of information and inferences about some aspect of the world — animal, vegetable, mineral or human, at a particular moment in time. It has been systematically *codified* and stored, so that it can be transmitted more or less faithfully from person to person. This means that it has to be made *explicit*, even though much laborious experience and tacit skill may be required to gain access to it and make use of it.

In its early years scientific knowledge simply grew by accretion. It was produced by observers, practitioners, inventors and thinkers who operated individually, and were seldom specifically employed to do so. In the modern world, however, scientific knowledge is typically generated deliberately, by collective action which is becoming what we call 'research'. So now, when we refer to 'science' we are usually thinking about the organised groups of people who perform this task professionally, as well as the particular type of knowledge that they produce.

Note, however, that this does not include the very much more numerous category of people who regularly use scientific knowledge in their daily work. In the case of medicine, for example, the skills of professional practitioners are based on systematic study of the accumulated wisdom of their craft. But only a few of them are actively engaged in extending, reforming or propagating this knowledge base. The vast growth and amazing potency of bio-medical research and of information technology, have given these branches of science immense influence on people's lives, but that does not mean that every physician or surgeon is a research scientist.

On the other hand, there are expert professions, such as architecture, where routine practice can often pose quite new problems. A novel design concept that solves such problems can then become a significant addition to the science that underpins the art. Indeed, it is sometimes maintained that every original human artefact or technique is a contribution to knowledge. That may be true in a very general sense, but misses the main point. In practice, it is impossible to 'reverse the engineering' and read out of, say, the latest ethical drug or mobile phone, the scientific knowledge that it embodies. The practical arts also involve experiential knowledge but that is too tacit — for doing rather than musing upon — to be formulated and transmitted to others.

So I need to go back here to a much older definition of a 'technology', which at the same time restricts it to the *explicit* or *codified* knowledge associated with a practical craft. Thus, I would *include* in science the divisions of Microsoft, Intel and Hewlett-Packard devoted to 'RDD&D' — Research, Development, Design and Demonstration — for their employees are certainly having to make explicit to one another the new technical knowledge they are acquiring as they work together on new products. But I would *not include* all the factories and shops where these so-called 'scientific' devices are manufactured and sold, nor the millions of people who use them with skill in their everyday lives.

However I see no reason for excluding from 'science' the production of knowledge about *people*. The conventional distinction between the 'natural' sciences and the 'social' sciences is extraordinarily difficult to justify theoretically. Needless to say, it vastly complicates matters when human beings are caught up in the loop. It is not at all so easy to make explicit the knowledge gained from a study of the 'moral sciences' or the 'historical sciences'.

But the social sciences are not just concerned with giving a convincing general account of what is going on in social life. Institutions such as bureaucratic governance, public finance and law are themselves powerful technical systems. Associated with them are great bodies of well-observed, closely argued scholarship, which must surely count as organised knowledge. Whatever the deficiencies of our basic understanding of management, economics, law, etc., these are 'human technologies' which are certainly no less scientific than, say, psychiatry or agriculture.

Indeed, agriculture provides us with a valuable metaphor. We no longer wander in the wilderness, gathering 'food for the mind' wherever we find it growing beside our path. We have learnt to get together, clear the land, cultivate the soil and sow our crops, with the expectation of a bountiful return. Only by the collective labour of highly-skilled groups, in family farms or in commercial plantations, could we produce the vast quantity and endless variety of mental sustenance required to maintain our civilization.

In that civilization, scraps of knowledge derived from science are scattered everywhere. But they do not always define themselves as such. They cannot be identified as examples of a single unique 'way of knowing'.[2] Nobody could doubt that we are referring to instances of a distinctive cultural form. But all that they have in common is that they each came originally out of the collective effort of an organised, expert group.

Science in society

Science, then, is not just any bit of knowledge labelled 'This is really true knowledge'. What makes it special is that it is the product of a special kind of social process. Of course that begs the question whether the pronouncements of any old club, sect or political party should be considered 'scientific'. Certainly not! As I have explained at length in previous books, there are significant conditions on the general type of group that we take seriously as a source of knowledge.

The way in which scientific knowledge is typically presented stems from the circumstances of its production. For example, it has to be *rational*. By that I don't mean that it has to be, say, based irrefutably on indisputable facts. I just mean that it has to be capable of being articulated intelligibly and communicated unambiguously from person to person within the producer group. This is the property that

[2]   Pickstone, J. 2000 *Ways of Knowing: A New History of Science, Technology and Medicine*. Manchester: Manchester University Press.

gives scientific knowledge so much rhetorical force.[3] Because it seems so reasonable we, the public, are easily persuaded that it is *reliable*. We feel bound to accept it as genuine knowledge, even though it may not be anywhere near as sound as it claims to be.

On the other hand, the expert communities within which scientific knowledge is thought out, produced and communicated, are typically highly *specialised*. What they regard as perfectly rational is all too often hidden from the lay person in a fog of jargon. In fact, the whole spectrum of 'organised knowledge' is minutely subdivided into innumerable specialties, speaking a Babel of mutually incomprehensible technical tongues. What is more, even within a single specialty, expert opinion is very seldom unanimous. All too often, competing individuals or groups give contrary answers to the same question. Streetwise citizens learn to move about freely in the contested spaces, picking and choosing for themselves the items of scientific knowledge on which they feel they are able to rely.

In practice, what we call 'scientific knowledge', even in the narrowest interpretation of the term, is so highly diversified as to seem almost incoherent. This diversity applies even more to the modern cultures where it has to find a place. This place is surely very different in a less-developed country such as Cambodia, Zimbabwe or Paraguay from what it is in Western Europe, North America or East Asia. The impact of modern science on a Christian, Buddhist or Hindu culture cannot be the same as on Muslim culture. It is not even quite the same in Spain as it is in France or Italy. Educational, religious, and even linguistic contexts have a significant influence on how the public receive its supposedly universal scientific findings. We are now beginning to realize that 'modernity' is not a single cultural tide sweeping away all other traditions, and that it is far from uniform in its scientific interactions with those other traditions. So a

---

[3]   Ziman, J. M. 1968 *Public Knowledge: The Social Dimension of Science.* Cambridge: Cambridge UP.

whole heap of reinterpretations of our scientific knowledge will be going on.

As we shall see later, this *pluralism* is very important in science; but that is not the present issue. The key point is that 'science' — and here I return to its broad definition as any form of organised knowledge — always seems peculiarly powerful because it comes to the public *already weighted with the social authority of the bodies that produce it.* It is not that it is so logically compelling or so efficacious in practice. Nor even that it is uttered by pundits, gurus or other particularly wise individuals. It is because it is issued in the name of a socially accredited *institution.*

Consider the position of an ordinary member of the general public. It takes little courage to question the incoherent assertions of a few self-satisfied or eccentric professors: it is quite another matter to defy openly the pronouncements of a whole university faculty, learned society or industrial research laboratory. When the views of many such bodies coincide, then their credibility becomes effectively unchallengeable, at least for the time being.

A major reason for the ever-growing public impact of science is that these bodies are becoming larger, stronger and more elaborately organised. From the seventeenth century onwards, of course, there has been continuous rapid growth in the number of people engaged full time in the production of knowledge. Laboratory and field research, technological development and design, and highly specialised academic scholarship, have all become regular professional careers. Every domain of theory and practice, from cosmology and theology to widget manufacture and certainly to district nursing, now has its own professors and research associates.

This intensification and proliferation of specialised personal roles is a commonplace feature of the modern world. What is more significant is that it required, and acquired, *organisation.* The advantages of intellectual teamwork and networking are not recent discoveries. Knowledge-producers, like other craft workers, found that they could operate more efficiently if they got together and divided the labour. Soon we had producer collectives, such as learned societies, professional colleges and institutes, and university departments. Academic entrepreneurs commonly establish around themselves whole 'schools' of graduate students, research assistants and associates. In due course, larger enterprises such as industrial firms and governmental bureaucracies, went into the business of *systematic knowledge production* in a big way.

Here is not a good place to discuss the actual processes by which all forms of science have been increasingly 'collectivized' in recent years.[4] All that needs to be said is that knowledge production is now largely in the hands of substantial undertakings, ranging from loosely coordinated academic institutions to tightly-managed commercial corporations and state agencies. Most scientific workers nowadays, whether engaged in basic or applied research, technological invention or design, or critical humanistic scholarship, are their salaried employees. They are hired 'do science' and can be fired (or at least shunted into much less attractive jobs) if they don't.

This is now the name of the game. What we call 'bench scientists', are mostly white collar workers, inhabiting big office buildings, and feeding together in self-service canteens. They are plugged into the internet through a private branch exchange or a local area network. They operate with immensely expensive apparatus, ranging from cute electro-mechanical devices to vast document archives and space satellites — none of which they personally own. Their findings are patented for profit, or published in journals and their books are marketed by large commercial corporations.

These are the material conditions under which scientific knowledge is generated in modern society. However individual its sources of inspiration, it is produced by societal *systems* dedicated to that purpose. I am not suggesting that this is an entirely modern development. It is nearly a thousand years since the first European universities began the

---

[4]   Ziman, J. M. 1995 *Of One Mind: The Collectivization of Science.* Woodbury NY: AIP Press.

collective pursuit and dissemination of learning in theology, law and medicine. Or should we go back to the academies and medical schools of ancient Greece? All advanced civilizations have provided a living for a few scholars — typically as priests and teachers — and aggregated them into what we should now call research institutes. But nowadays we produce scientific knowledge as industriously, and on the same scale of effort, as we produce, say, our clothes and furniture.

This is too well-known to require detailed documentation. Consider simply the financial dimension. For some 25 years, national governments have been issuing official 'indicators' showing how much is being spent, in the public and private sectors of the economy, on scientific and technological research and development. To this should be added a substantial sum for the non-routine knowledge that is being systematically accumulated in other spheres of life, such as law and public administration. For a country that counts itself as economically developed, the total normally amounts to several per cent of the gross domestic product.

The physical scale of this effort is not directly visible because it is distributed so widely. The knowledge production industry is not concentrated into large factories like the manufacture of motor vehicles, nor is it dominated by an oligarchy of self-advertising corporations like supermarket chains. Some high-tech companies and government departments have very large research laboratories, but these each employ no more than a few thousand researchers and technical staff. Very often, one has difficulty in finding them in their inconspicuous buildings in a corner of the main company site, or in a leafy science park attached to a distant university campus. Even a 'big science' facility such as a particle accelerator does not loom larger in the environment than a steel mill. Although a modern university may have an annual research turnover running into many hundreds of millions of pounds, this is typically subdivided amongst numerous nearly independent departments, institutes, research centres, and so on.

In practice, however, for all the talk about small enterprises funded by venture capital, it is not so easy to make a
living just by producing knowledge and marketing it as
'intellectual property'. Much of this effort is held together
by financial, bureaucratic and contractual bonds. These
networks are often much more extensive and influential
than they seem on formal organisational charts. For example,
in the pharmaceutical industry, the research 'boutiques'
created by scientific 'entrepreneurs' very frequently operate as specialist knowledge providers for major companies.
Many academic institutes that are nominally autonomous
are actually dependent on a sustained flow of project
money from the private and public sectors of the economy.
(As we shall see, these commercial and political influences
are growing in strength, and are having a profound effect
on the nature of the science that they foster.)

Science is also held together by intellectual, technical and
methodological linkages. The knowledge produced by each
group, whether big or small, is very largely compounded of
knowledge obtained from somewhere else — from a variety
of other quite separate groups who have published it in one
way or another. Indeed, the concept of science as 'public
knowledge' is based on the formal and informal networks
through which it is communicated.[5] In practice even the
linkages are highly specialised and are not connected into
an orderly structure. Even in its narrowest sense, 'science'
does not constitute a unified body of knowledge. In a wider
perspective, it may seem an amorphous mass of reliable
facts and instructive theories, over-confident techniques
and contested inferences. in which many quite distinct
'sciences' are lumped together higgledy-piggledy without
any overall plan.

It is quite misleading, therefore, to refer to 'science' as if it
were a unitary social or intellectual enterprise. It is not like
a church promulgating a specific doctrine or a public
company selling a well-defined product. It has no Supreme

[5]   Ziman, J. M. 1968 *Public Knowledge: The Social Dimension of Science.*
Cambridge: Cambridge UP.

Pontiff to lay down the Law on a disputed issue, nor a Board of Directors to approve a corporate plan. It speaks with a multitude of voices, often quite contradictory.

On the other hand, modern science is not really just a voluntary community, an autonomous 'republic' within the general polity.[6] It is socially fragmented, but not into isolated individuals — and many of the 'fragments' are actually large, tightly-managed organisations. It is required to respond to market forces or to serve the nation-state, and much of it is subject to external oversight and strategic control. Metaphorically speaking, its peasant holdings are being enclosed into manorial estates, and its cottage industries are being consolidated into factories.

Nevertheless, despite its intellectual and organisational incoherence, science has an established place in modern society. The systematic production of knowledge is a highly-regarded professional activity in which numerous very talented people make well-rewarded personal careers. It attracts large financial investments from hard-headed industrialists and public officials. Its leaders are near to the heart of economic and political affairs. Its products are eagerly sought after, lauded to the skies, and carry very great weight even in the most intimate aspects of our lives. By every tangible or intangible measure it is one of the central institutions of our culture.

Science as politics

And yet people may still ask themselves: 'What is science *for?*' It clearly plays a very important role in society, and different sections of science can be seen most clearly by people with different interests. Young mothers, for example, will be most concerned for the welfare of their little ones and for them the chief role of science becomes that of health manager. So if we ask again — what exactly is the role of science? — it becomes easier to pinpoint particular responsibilities for which science is becoming the load-bearer. Most people

---

[6]　Polanyi, M. 1962 The republic of science: its political and economic theory. *Minerva* 1, 54-73.

don't have any difficulty in answering questions like these about, say, the police or the brewing industry. But science itself is more difficult to evaluate. It is remote, uncontrolled and yet immensely powerful. So they feel they need answers to these questions before they can decide properly whether they want to support it, or believe in it, or put themselves under its control, or even attempt to do without it.

Unfortunately, this is not the sort of question to which there is a scientific answer. Scientific experts in each field of knowledge will naturally urge everybody to support and believe in *them* — even to the point of meekly submitting to their entirely beneficial rule. But those same experts are most unlikely to advise an equally favourable attitude towards the experts in *other* fields, especially where the human sciences are involved. Indeed, the scholars who actually specialise in 'science and technology studies' seldom pose such grand questions directly, and are mostly canny enough not to try to answer them. All that can be read between the lines of their writings is that the natural scientists who still insist that SCIENCE is an unmitigated, universal good have somehow found themselves spouting what is clearly metaphysical fiddlesticks, a conclusion with which I would thoroughly agree.

On the other hand, this question should not be left quite open. Of course one can always say that society itself determines the place it gives to science. In effect, it is just a matter of accepting actual cultural practice. Science is as science does, and what it does is always within a specific social context.[7] Its 'role' then is simply to perform as is normally to be expected of it, in the ordinary way of things.

This fatalistic attitude may well be appropriate in relation to deeply embedded cultural forms like marriage, or organ-

---

[7]   Barnes, B. & Edge, D. (ed.) 1982 *Science in Context: Readings in the sociology of science*. Milton Keynes: Open UP, Ezrahi, Y. 1980 Utopian and pragmatic rationalism: the political context of scientific advice. *Minerva* 18, 110-31, Knorr-Cetina, K. D. 1981 *The Manufacture of Knowledge: An Essay on the Constructivist and Contextual Nature of Science*. Oxford: Pergamon, Wynne, B. 1991 Knowledges in context. Science, *Technology & Human Values* 16, 111-121.

ised crime. But it seems to dodge the issue when applied to an institution that is so highly organised, and attracts so much purposeful effort, as modern science. As we have seen, many shrewd people are willing, individually and collectively, to 'do' science, or fund others handsomely to do it for them. They must surely have some idea of what they, and their fellow citizens, are going to get out of it?

So science is not 'a wind that bloweth where it listeth': it is generated and constrained by human intent. Instead of continually sampling the views of the general public on science as *received*, researchers ought to focus for a change on the actual purposes for which it is *produced*. Imagine then the questions that would have to be asked: 'Tell me, Sir, why your company has just invested a hundred million pounds in a DNA sequencing laboratory?'; 'Why, Madam, did you vote in Congress for another half a billion dollars for the Space Station?'; 'What can be expected from the Institute of Legal Ethics for which you have donated ten million Euros to the University?'; 'Is your Ministry satisfied with the research programme for your new Submarine Exploration Platform?'; 'Why doesn't the budget for the Museum of Antiquities make enough provision for its research activities?'; 'What were the arguments leading the Academic Senate to close down the Faculty of Theology?'; 'How does your Committee justify the grants it is giving in the field of Mediaeval Archaeology?'; and so on.

It is not difficult to guess at some of the answers that might be given to such questions. But for the moment let us note to whom they would be addressed. Informed replies could only come, of course, from 'decision makers' in the state bureaucracy, business and academia. They relate to operations in the *political* dimension of society, especially in its *commercial* sectors. Science is a major element of the *economy*, and must adapt itself to the requirements of the *polity*. Scientific knowledge is so vital to the health, wealth and happiness of the nation that its production is of active concern to many of the powers that be. It is also very expensive — and this concerns whose who need to know where the money comes from. So modern science is very largely driven and shaped by these powers — governmental, financial, industrial, military, clerical, legal, and so on.

As a consequence, every advanced society prescribes for science a public role that is consistent with the predominant political agenda. The means for the production of knowledge are an indispensable element of the whole social system, and must therefore be seen to conform to that system. I am not suggesting, as a doctrinaire Marxist might have put it, that science is merely a component of the ideological superstructure imposed by the ruling class (or something of that sort). I am just saying that science is now so clearly one of the major sources of social power that its outward form and supposed function are largely governed by whatever force, group, idea or person lays claim to this power.

One should not, of course, imagine that science actually performs according to its prescribed script. Indeed, this script is mostly wishful thinking, constructed by a powerful elite to justify its position. Human affairs seldom follow the agendas ordained for them. Even the most dominant social groups do not get the world they want. Science, in all its forms, is a dynamic activity. It produces many wild cards in the political game. Many other social influences are at work on, with and around it. For example, the frequent obstinate resistance of ordinary people to its knowledge claims does not fit well into the official scenarios.

It is frequently argued, however, that the history of science, technology and other cultural forms is mainly determined by political and economic forces. These forces are largely hidden, but they sometimes surface as institutional voices which assign to science both a public image and social role. These voices exert an immense influence on the actual place of science in society, and should not be ignored just because, due to the rising costs or the increasing risks of projected outcomes, they do not always report completely reliably what is actually going on.

Some political agendas for science

As we have seen, scientific knowledge is not ordinarily encountered as a collection of facts and theories that just speak for themselves. It comes into the public realm as the product of the collective endeavours of warranted experts. These warrants are signed by technical communities, but they are issued under the authority of other powerful social institutions which are the governors of science. Their wording indicates the purposes for which this knowledge was produced and for whom it is certified for use. The social function of science is thereby publicly proclaimed.

Let us look at some examples of the different roles that science plays in different types of society. Consider, first, its place in a 'traditional' society, a term that covers a miscellany of social systems ranging from simple hunter-gatherer cultures to sophisticated agricultural empires. What such societies have in common is that the production of knowledge was not an organised social activity. What we now call 'science' was not differentiated from other sources of practical or theoretical expertise.

Human societies have always had individuals who were recognized as outstandingly skilled in particular practices, and who shared their learning with their kinsfolk or with fellow members of their guild. But in a traditional culture they gained their living as practitioners and masters of apprentices, not as scholars or inventors. Similarly the generation and consolidation of new knowledge was incidental to the ongoing flow of social life. The idea of codifying and extending it was simply not on the political agenda. Typically, those who now seek to revive the authority of ancestral traditions renounced the aid of orderly intellectual enquiry. They may present their racial, paganistic, mystical or divinely revealed notions as perfectly rational. But they are often unwilling to open them to debate in philosophical, theological or other systematically 'scientific' terms.

By contrast, a society that is self-consciously 'theocratic' is bound to cultivate detailed discussion of the doctrines to which it adheres. Colleges of experts and archives of arcane knowledge luxuriate around their ethical and legal implications. As issues of principle are debated and codified they become the basis for an established 'science'. On their own territory, Buddhist monks, Muslim mullahs, Talmudic rabbis and Scholastic clerics could dispute just as productively as any bunch of Marxist philosophers, neo-classical economists or theoretical cosmologists.

But the substance of this knowledge was often kept secret from the 'laity' and did not provide much guidance to the problems of everyday life. The technical sciences that began to evolve in practitioner communities were often treated with disdain and were never allowed to challenge the higher levels of learning. In such a theocracy, the official role of all forms of scientific thinking is to sustain the authority of the established religion — and, of course, of its priestly elite. The fate of Galileo perfectly illustrates this answer to the question 'what is science for?' It is to sustain a picture of the frontiers of scientific knowledge as prescribed in the books of one or other set of the religious elite. Present-day fundamentalist creationism, or 'intelligent evolution', shows that it is still being assigned this role by people who are about today.

Of course the growth of the natural sciences and their associated technologies since the seventeenth century has given them much more political weight today. In addition to proclaiming the glory of the heavenly King, they advertised the benevolent patronage of His regents on earth. Autocratic rulers began to point to scientific progress as evidence of their benefits to the nation. They gained cheap political Brownie points by officially fostering the production of knowledge. And when eventually they fell from power, the 'royal' academies that they had founded were reborn as 'national' institutions, thus continuing to announce that the true role of science was to serve the nation-state.

Indeed, in the extreme case of Soviet Communism, the official state doctrine — 'Marx-Leninism' — was supposed to be founded on 'scientific materialism' and all scientific activity was incorporated, formally and informally, in the state apparatus. Not only did this mean, as shown in the case of Lysenko, that scientific knowledge would not be allowed to come into conflict with the tenets of the ruling ideology. It also meant that the organisations and technical communities that produced this knowledge were quite powerless to defend their findings against political attack. In effect, the only permissible public attitude towards science could be stated simply: 'Our People are all for Science: Our Science is for the People'. The contradiction between this slogan and reality was kept as a state secret.

More naively, some advocates of 'scientific socialism' believed in 'technocracy'. Writers such as H. G. Wells, J. D. Bernal and C. P. Snow held that 'science', by which they meant the natural sciences and their associated technologies, should be the prime source of authority in society. They envisaged a social system run on entirely rational lines, where normal politics had been eliminated. The public should thus look up to science — and to scientists, of course — as the sole centre of social decision and action. Fortunately, no such system has ever been put into practice!

What we have got now, though, is 'capitalism'. Here, all social action is supposed to be in the hands of private enterprises, that is, of commercial firms competing freely for customers in 'the market-place'. Scientific research and technological innovation are merged into *technoscience*, a widely dispersed activity undertaken by independent corporations seeking to profit from the knowledge thus produced. By focussing on the notion of *consumer choice*, this system avoids being openly technocratic. But it subordinates the social sciences to neo-classical economic doctrines that purport to prove that market freedom will automatically optimise human welfare. Thus, countries such as Singapore and Korea justify state expenditure on science as an 'investment' that will eventually produce handsome dividends in national competitiveness and wealth creation.

But capitalism still has its critics. Many people nowadays, including many scientists, actively resist the pervasive influence of the multinational companies and their political allies. They agitate for an alternative social system where this power has been eliminated or drastically curbed. In such a utopian vision, technoscience would have been set free of its governmental and corporate masters and would operate as an empowering force for popular liberation, sustainable development, and so on. It would be quite wrong to say that all 'anticapitalists' are opposed to science in general. But they do insist that its present societal role is highly suspect and often extremely antisocial. Although they seldom say so in so many words, they seem to believe that the scientific enterprise as a whole would need to be 'politically corrected' before it could it carry out its social functions in a truly enlightened spirit. But despite the fears of many self- appointed defenders of science,[8] this radical agenda is still far from being accepted by those occupying the seats of political power.

I am not pretending to offer a serious sociological or historical analysis of how various societies, real or imagined, have actually related themselves to science. The social systems that I have sketched out are obviously hypothetical or highly schematized. But they do indicate some of the diversity of the services that science is officially called on to perform. Although some of these are very dubious ideological constructs, this type of public discourse has real effects. Nowadays, when the organised production of scientific knowledge is so costly, and employs such quantities of talented labour, it greatly matters what people think it is for.

And yet, as we shall see, science performs a variety of other societal functions that are seldom mentioned in political speeches, commencement addresses, or revolutionary tracts. Its place in the modern social order has become so familiar that we tend to forget how important it is. Like democracy and equity, we take its benefits for granted, neglecting the institutions through which these are maintained.

---

[8]   Gross, P. & Levitt, N. 1994 *Higher Superstition: The Academic Left and its Quarrels with Science*. Baltimore MD: Johns Hopkins UP, Gross, P. M., Levitt, N. & Lewis, M. W. (ed.) 1996 *The Flight from Reason*. New York NY: New York Academy of Sciences, Holton, G. 1993 *Science and Anti-Science*. Cambridge MA: Harvard UP.

Science in a pluralistic polity

Most readers of this book, however, will probably have found the previous section rather irrelevant. Very likely they live in an economically developed region of the world. That is, they live in a society that does not conform to any of the above stereotypes. So the role assigned to science in the local political agenda will also be somewhat different.

Consider, for example, current concerns about 'public attitudes to, or awareness of, science'. This implies that there is such a thing as a 'public' capable of having diversity of such attitudes. This would have been meaningless in a traditional, theocratic, totalitarian or technocratic society. In such a society, the views of people in general don't count. If science can be said to exist there at all, it is as an instrument of government, or at least a loyal ally of the ruling elite. Again, market capitalism treats people solely as 'consumers', with only a limited interest in the production of knowledge as such, whilst anti-capitalists are likely to be anti-scientists as well as 'anti-consumers'. A genuine 'public' can only operate in a free, open, democratic polity, where a number of respected social institutions jostle visibly for votes, voices, or trade.

In other words, the standard accounts of 'science in society' assume that we are concerned with a *pluralistic* society where science itself is just one of these competing institutions. This produces a plurality of public attitudes, not only because science is encountered under a wide range of circumstances but also because it is in the service of a plurality of political agendas. We count ourselves as fortunate to be able to live in a society where there is no central authority or ideology capable of prescribing a unique social role for the natural and human sciences and their associated technologies.

All that I take to be generally agreed. Indeed, it can be argued that such basic political rights as freedom of speech and assembly are essential conditions for the production of reliable scientific knowledge.[9] It can also be argued that the whole polity ought to be modelled on 'the Republic of Science' — that is, on the way that academic scientific communities govern themselves.[10] But in this present work I want to turn these arguments on their heads. I propose to show that democratic pluralism itself has come to rely on science for its own continued existence as a functioning political system.

In brief, the overarching ideology of our political culture is its toleration of diversity. That means that it often has to reconcile a variety of competing interests and demands. This requires the active participation of well-informed people and institutions who can be trusted to operate reasonably impartially at the points of conflict. Academic science — and here let me re-iterate that I use the term in its widest sense — has customarily fulfilled this need.

Thus, the diversity of public attitudes towards science is not, somehow, a weakness. It should be welcomed as a healthy indication that science as a whole is not identified with any particular interest group or cause. People are uncertain about 'the place of science in society' precisely because it has many different 'places' in the social structure, and so many different parts to play in the ongoing social drama. Our pluralistic society is itself stabilized by this plurality of the whole scientific enterprise.[11] The social order is stitched together by threads of meaning and action generated by this inchoate but many-splendoured endeavour.

We should bear in mind another world of pluralism in science which has only recently emerged from the shadows. Does the pluralism of science apply also to the knowledge and culture of those who, while often calling themselves 'non-scientists', in practice feel equally at home in science,

[9]   Ziman, J. M., Sieghart, P. & Humphrey, J. 1986 *The World of Science and the Rule of Law*. Oxford: Oxford UP.

[10]   Polanyi, M. 1962 The republic of science: its political and economic theory. *Minerva* 1, 54-73.

[11]   Ziman, J. & Midgley, M. 2001 Pluralism in science: a statement. *Interdisciplinary Science Reviews* 26, 153.

and vice versa. This broadens the readership of science and makes it more accessible to both groups.

> *Scientific knowledge is incurably pluralistic. It requires systematic study of the world at every level of complexity ... but the present-day views of science are unhealthily polarised. Those who assert that science can completely discover reality are opposed by those who insist that scientific knowledge is simply a social construct ...*
>
> *In practice however most working scientists take neither of these extreme positions. They are aware that established procedures ultimately rely on human judgements, but are confident that they produce peculiarly credible and reliable bodies of knowledge.*

Ziman and Midgley 2001.

TWO

Instrumental and Pre-Instrumental Roles

Technoscience

By any measure, the predominant mode of knowledge production in modern society is *technoscience*. What used to be distinct scientific disciplines have merged with cognate technological disciplines to form a seamless enterprise. The knowledge produced by this enterprise ranges from abstract theory to real-world practice. At one end it consists solely of intangible ideas conveyed through papers and books: at the other end, it is embodied in useful artefacts and techniques. At one end it is a theorem or a model: at the other end it is a blueprint or a prescription. The two end forms are clearly totally different and were formerly completely disconnected. Nowadays they are linked by a chain of intermediate modes of knowledge production.

So we can now see all this as fundamental research, basic research, applied research, technological development, product design and prototype testing. The various categories of scientific work grade imperceptibly into one another, right across the range.

Take, for example, 'biomedicine'. Its very name indicates that it combines the traditional basic science of biology with the traditional technical practice of medicine. Where now does one component stop and the other begin? They interpenetrate and interweave. On Mondays, Wednesdays and

Fridays a scientist is trying to make fundamental discoveries: on Tuesdays, Thursdays and Saturdays (working overtime, but not being paid extra for it!) the same scientist may be making useful inventions: on Sundays, the same scientist is beginning to worry that these three modes of discovery are very closely connected.

Imagine being put down in an unnamed research laboratory. Observe the instruments and experimental techniques — test-tubes, agar dishes, microscopes, centrifuges, incubators, refrigerators, etc., etc. Look at the terms, formulae, graphs and diagrams in printed texts and notebooks, on whiteboards and computer screens. Listen to the words and arguments that fill the air. Notice the patent application form in the same box file or computer 'folder' as the draft of a Letter to *Nature*. What they are doing is human genomics. But is this the Laboratory of Molecular Biology at Cambridge, where the structure of DNA was discovered? Or are you in the Science Park on the other side of town, in a newly-formed, spin-off, venture capital, pharmaceutical enterprise, hoping to put on the market a wonder drug for cancer? How long will it be before somebody finally lets the cat out of the bag by mentioning a possible Nobel Prize for a friend or, alternatively, that the rating of a company on the Stock Exchange is going through the ceiling.

It is not just that the same data bases, the same theoretical paradigms and research techniques, the same highly educated and expert personnel, are being used right across the board. It is that technoscience presents itself as a unitary enterprise, devoted systematically to the production of its most tangible outcome — technological innovation. But the essence of a technology is that is about *practice*. It deals with the ways in which people act intentionally to achieve their material ends. It codifies or produces the knowledge that people need in order to reach these ends. It is a *tool*, an *instrument* for practical human action. Technoscience is thus cultivated primarily for its *instrumental* capabilities.[1]

---

[1] See further the chapters that follow, with the diagram 'But Research Cultures Are Changing' (p. 115).

The instrumental function

The Frascati Manual,[2] proposing a standard differentiation of aspects of scientific research and development (R&D), describes three related activities:

- **basic research** is experimental or theoretical work undertaken primarily to acquire new knowledge of the underlying foundation of phenomena and the observable facts, without any particular application or use in view;

- **applied research** is also original investigation undertaken in order to acquire new knowledge. It is, however, directed primarily towards a specific practical aim or objective; and

- **experimental development** is systematic work drawing on existing knowledge gained from research and practical experience that is directed to producing new materials, products or devices; to installing new processes, systems or services; or to improving substantially those already produced or installed.

The boundary between basic and applied aspects of R&D is notoriously subjective, prompting some commentators to combine these categories into 'strategic research', and then subdivide **basic research** into 'pure-basic' and 'orientated-basic' and **applied research** into 'strategic-applied' and 'specific-applied':

- **Pure-basic research** is carried out for the advancement of knowledge, without working for long-term economic or social benefits, and with no positive efforts being made to apply the results to practical problems or to transfer the results to sectors responsible for its application (often called 'blue skies' research;

- **Orientated-basic research** is carried out with the expectation that it will produce a broad base of knowledge likely to form the background to the solution of recognised or expected current or future problems or possibilities;

---

[2]   *Frascati Manual 2002: Proposed Standard Practice for Surveys on Research and Experimental Development*, published by the Organisation for Economic Co-operation and Development (OECD). This section draws on material at http://www.dti.gov.uk/science/.

- **Strategic-applied research** is directed toward practical aims, but has not yet advanced to the stage where eventual applications can be clearly specified; and

- **Specific-applied research** will have quite specific and detailed products, processes, systems, etc. as its aims.

The wider term 'strategic research' describes work that has evolved from pure-basic research and where practical applications are likely and feasible but cannot yet be specified, or where the accumulation of underlying technological knowledge will serve many diverse purposes.

In their naïve sketch maps of the social order, modern political and economic elites equate all science with instrumental technoscience. They thus proclaim that its sole function is to guide and inform practical life. In their rhetoric it is assigned an entirely instrumental role. They point out, for example, that a significant proportion of the output of scientific research has already been put to practical use, or has quite foreseeable applications. Hence they deduce that further knowledge produced specifically for this purpose could surely benefit any cause that seeks it and exploits it diligently. So they publicly foster science as a powerful means of achieving the material goals that they happen to consider the most desirable.

This line of thought has commended science to the ruling groups in all forms of civilised society. Even in a theocracy, it is used to justify the maintenance of colleges of learned clerics, whose collective expertise on heavenly matters—for example, astrology—is considered indispensable to the earthly welfare of the state. It is the basic ideology of every would-be technocrat; even their utopian opponents act as if it were also their credo. 'Judge science by its utility', they all say, 'and value it pragmatically in terms of what it gives you, for good or ill.'

Nowhere is this reasoning more firmly entrenched and publicly reiterated than in our modern, free-market democracies. In the country of the Smiths, the principal national objective is to keep up with Jonesland in economic growth and development. So the role of science is to contribute to 'wealth creation', 'industrial innovation' and 'international

competitiveness'. We are also anxious about 'national security', 'public health', 'social welfare', 'crime and punish-
ment', and so on. The call for the scientific knowledge will
surely enhance these as well. Such are the values we live by,
and they also are what our science is for.

The voices of government, industry, the media, political
parties of all persuasions, and most members of the public,
sing different songs, but all in much the same key. They
confound science with technology, and celebrate the
technoscience that seemingly makes all things possible,
including the cure of the ills that it has itself created. If there
is such a thing as a 'contract between science and society',
then these are what science has signed up to provide a boost
for, if it is required.

The pre-instrumental function

I have been 'for' science since childhood, not least for its
magical capabilities in a bewitching world. I am certainly
not disputing its fantastic utility as a social institution. That
is not my target, here or anywhere else in this book. All I
want to point out that it also performs a number of other
very important and complexly interwoven social and
humanistic roles.

Let me emphasize, then, that I am in no way denying that
science, both in general and in particular, is immensely useful. When faced with almost any practical problem, it is the
first tool we should pick up and apply to the job. Its results
and methods are usually worth trying, even if sometimes
the solutions they offer don't seem to work at all, or need to
be treated with great caution. I cannot accept the notion that
one should completely renounce the use of such a potent
instrument as science just in case it might occasionally lead
one along the wrong path.

In my opinion, scientific knowledge has indeed been the
prime source of enormous human benefits. The belief that it
has played a vital social role in producing these benefits is
entirely justified. The fact that science has also produced a
number of very serious disbenefits does not controvert this

belief. It merely demonstrates that science is at work in a new and difficult field. It is not easy and sometimes proves to be a somewhat *blunt* instrument, and emphatically not one of those smart self guiding weapons of attack which the more utopian technocrats imagine so fondly to themselves. Science often presents us with hard problems, and I believe most scientists enjoy it being like that. What would be the fun of pursuing too simple a problem where the solution just drops like Newton's apple, into our laps?

Thoughtful technocrats also understand this very well. They fully recognise that research is a venture into the unknown, that its stated objectives are speculative, that they are seldom actually reached, and that even then they may not turn out to be on entirely firm ground. The true utility of the knowledge one has gained may only be realised in hindsight. Tons of gravel may have to be washed away to win a few ounces of real gold.

The fact is that most of the knowledge produced by scientific investigation and technological invention does not have any obvious practical use. Nevertheless, its *potential* utility is so enormous that it justifies the expense of obtaining it. Like prospecting for gold, it's a gamble; but history has shown that such bets pay off handsomely, over all.[3] What is more, this is not a random lottery. With a modest expenditure of technical imagination, it is often possible to construct plausible scenarios where the most likely results of the research might quite conceivably be exploited, in industry, agriculture, war or medicine. So the punter can use his head to shorten the odds against his own failure by using new ways of thinking about the problem which may be beginning to show some success. No wonder then that they so often push on hopefully along the same path, sometimes persevering longer than his/her manager would really like.

In the world of science policy, this is the rationale for undertaking 'strategic research'. Scientists like this term, for

---

[3]   Mansfield, E. 1991 The social rate of return from academic research. *Research Policy*.

it is wonderfully elastic. Its anchor points are conjectural. So the project proposal can be twisted and stretched to cover all sorts of hypothetical outcomes, of varying degrees of plausibility. Or it can promise many intermediate benefits, such as strengthening the theoretical base, discovering applicable principles, developing novel techniques, extending technical capabilities, 'spinning-off' serendipitous innovations, etc. In effect, a strategic research programme is a scatter-gun, not a rifle. (Aim it roughly in the right direction, pull the trigger and hope to hit some juicy targets!)

But near the bottom line remains the hope of final practical gain. We can all quote Benjamin Franklin's cliché about invention—'What is the use of a new-born child?' The key word is *use*. That is what an invention is for, and that is the potentiality that we seek in other forms of scientific knowledge. Even if its practical value is not immediately obvious and may never eventuate, it is a possibility to be taken seriously and assessed accordingly. Although not overtly utilitarian, this mode of knowledge production has 'pre-instrumental' motives that are integral to its overall instrumental social role.

Who then calls the tunes?

On this view, then, is science is just a means for achieving the purposes of its masters and for delighting its practitioners? Thus, in totalitarian hands, it can become an instrument of oppression. That was the chilling message of George Orwell's *1984*, and much other science fiction. In such a society, the role of science is not only prescribed in the predominant political agenda: the scripts themselves are drafted in advance to favour the powers that be. Knowledge production is reduced to political action by other means. What more is there to say?

However, in a democratic, mixed economy, power is divided and contested. Technoscience is the creature of both government and industry, working more or less independently or in loose, somewhat uneasy partnership. The public institutions and commercial firms that fund research and control its products are mostly large and highly influential. But their technocratic tendencies are frustrated by political and economic rivalry. As independent constituents of a competitive, pluralistic society they may well have conflicting missions and agendas, and they often have no collective vision of 'the place of science in society'. They simply deploy their technoscientific capabilities and extend their technical resources, win markets, regulate or sue one another, gain public approval or otherwise and further their particular interests. In their day-to-day encounters with the public, they take it for granted that that this is all that science is for.

This mundane diversity of goals and roles in the private sector is not, of course, inconsistent with public rhetoric about fostering science collectively for wealth creation, etc. It is the duty of a modern government to set national economic objectives, over and above the conflicting objectives of individual enterprises. So they earnestly endeavour to hitch science also to those more distant stars.

From this perspective, the role of scientific knowledge becomes synonymous with its particular uses. A free society is home to a plurality of political and economic actors, at various levels. It thus generates a plurality of goals for scientific activity, on the large or small scale. The 'external' sociology of science is fixated on the organisations and institutions that envisage, seek, dispute, pervert, facilitate, misconceive and even control these goals, and employ science to achieve this. All too often, the science itself is lost from sight amidst the conflict of larger social forces.

In effect, all discourse about science then becomes political and managerial. It is expressed in words like 'policy', 'priorities', 'plans', 'problems', 'programmes', 'projects', 'performance', etc.[4] In conformity with the prescription for 'Mode 2' of knowledge production, it deals with 'problems

---

[4]   Ziman, J. M. 1994 *Prometheus Bound: Science in a Dynamic Steady State.* Cambridge: Cambridge UP, Ziman, J. M. 1995 *Of One Mind: The Collectivization of Science.* Woodbury NY: AIP Press.

arising in contexts of application'.[5] So instead of being differentiated into academic disciplines, it is disaggregated into practical 'problem areas'. What then counts most is who 'owns' each problem, to whom are its solutions of material interest, who has the means to seek them, and how well they succeed.

Most scientists are happy enough to be employed thus. All that they ask for is respect for their technical capabilities and liberty to exercise these in their daily work.[6] Apart from a few technocratic dreamers, they agree with Winston Churchill that scientists should not aspire to be 'on top'. But many are uneasy at his punning counterpart that they should always be 'on tap'. Are their hard-won discoveries, inventions, data, concepts, designs, and techniques just a stream that can be turned on or off like a fire hydrant? Does the tradition of *academic freedom* count for nothing? Where is their much-vaunted *community* in this picture? Has the world changed so much that the notion of scientific and scholarly *autonomy* is now quite obsolete?

To put it another way: scientists themselves are beginning to wonder what their labour is really *for*. In particular, why was it publicly supported as a valuable social institution in the days before it came under all those instrumental thumbs? They protest strongly that it ought to continue in that quasi-autonomous mode. But now they find it increasingly difficult to put up a defence against the strictly utilitarian reasoning that prevails in many influential layers of society at large.

This vulnerability is felt most keenly by scientists in fields such as cosmology, pure mathematics, ancient history — or even the philosophy of science. For some good reason or another less obvious one, they are still being given public

[5]   Gibbons, M., Limoges, C., Nowotny, H., Schwartzmann, S., Scott, P. & Trow, M. 1994 *The New Production of Knowledge*. London: Sage, Nowotny, H., Scott, P. & Gibbons, M. 2001 *Re-Thinking Science: Knowledge and the Public in an Age of Uncertainty*. Cambridge: Polity Press.

[6]   Jagtenberg, T. 1983 *The Social Construction of Science*. Dordrecht: Reidel.

funds to go on producing knowledge for which there is no conceivable material use. How can they possibly justify this expense? How can they compete with more practical disciplines such as materials science, or agricultural economics? At best they concoct an implausible pre-instrumental rationale—'who knows, we might find a new source of energy, or a quick fix to Stock Market crashes, or the political wisdom required for international peace?'. Pressed hard, they talk vaguely about 'cultural values' or 'blue skies research', or repeat endlessly a mantra about the seeking knowledge 'for its own sake'. For all their bravado and Nobel Prizes, they can't help feeling that they belong to a past age.

Meanwhile at the top of the political tree the most simplistic view of valuable new scientific discoveries dropping out of the sky continues to lead our masters astray. It is in education that this is to be seen at its worst. *'We need more pupils to take the hard sciences at school'* they continue to murmur to each other, especially when the nation's books will not balance. If the country is in a bad way the *pupils need to be taught to enjoy science more* as though it can just happen regardless of the fact that few of the MPs have enjoyed learning science during their own education. And so, it follows (doesn't it?) that they will not be making those valuable discoveries which will bring our country more wealth to go into balancing the Chancellor's budget.

These different problems arise, I would argue, because the social function of science is thought to be primarily utilitarian. The political and economic ideologies of our era are blind to its *non-instrumental* roles. And yet, as I shall endeavour to show, these roles are important in our pluralistic, multicultural, and considerably disturbed Britain, in other important ways. So this will be the subject of our next chapter.

THREE

The Non-Instrumental Roles of Science

Images, attitudes, people

It would be silly to suppose, of course, that all scientific knowledge nowadays is being produced for practical use. The spirit of *pure* science is still very much alive. It dwells for example in the immense particle accelerator facilities at CERN, near Geneva, and at Fermilab near Chicago. It has smaller shrines in every university laboratory and library. Its products are exchanged in vast bulk on the internet. It is celebrated annually, world-wide, in the Nobel Prize award ceremonies. Billions of dollars are sacrificed publicly to it, in the budgets of every advanced nation.

But the conventional justification for this dedication to the Goddess of Truth is hollow. It is dressed up to conform outwardly to the utilitarian ideology of out times. Stripped of pre-instrumental fantasies it is seldom more than a metaphysical belief in the absolute value of 'knowledge' as such. I am not in the business here of either affirming or disputing this belief. As I have argued at length elsewhere, what we call 'truth', even in the strictest scientific sense, is both 'discovered' and 'constructed'.[1] In any case, this line of argument deserves a lot more careful thought than it usually gets.

What I am getting at is that the proponents of 'pure' science usually overlook the strongest card in their hand—

[1]   Ziman, J. M. 2000 *Real Science: What it is and what it means*. Cambridge: Cambridge UP.

its beneficial *non-instrumental* social roles. People are undoubtedly aware that the production of scientific knowledge brings many intangible benefits, and some serious costs to many different parts of society. But somehow these items never seem to figure on the balance sheet.

As we shall see, these benefits are quite varied. In an open, democratic, pluralistic society, we rely on science for instructive and sometimes frightening *world pictures*, for critical *rational attitudes* and for reliable, independent *expertise*. Why are such obvious boons not more discussed? Perhaps we are all so bemused by the practical achievements of technoscience and its managers that we just take these other functions for granted. And yet, without them, our civilisation would not be viable. Science is a Pandora's Box of technocratic temptations: it is also an armoury of defences against obscurantism and tyranny.

Unknotting the oxymoron

But before we go further, we need to clear some linguistic space around the concept of a 'non-instrumental role'. Is this not an oxymoron—a self-contradiction? Doesn't the notion of a 'role' imply a purpose? Shouldn't it then be considered an 'instrument' for achieving that purpose? So we are apparently talking about a non-instrumental instrument for achieving a purpose-less purpose. Such a nonsensical entity slips out of our mental grasp.

Perhaps we can gain some purchase on it by replacing the metaphorical term 'role' by the more formal word 'function'. Sometimes, echoing the title of Bernal's famously influential tract,[2] I am inclined that way. I am well aware that this word too raises objections, perhaps because it hints at the notion of an 'intelligent design' in a non-religious sense. In biology, for example, the 'function' of an organ is to contribute to the operations of the organism as a whole. It implies that this organ is an active component of a 'system' with a pre-ordained ongoing collective existence. Indeed,

---

[2]   Bernal, J. D. 1939 *The Social Function of Science*. London: Routledge.

'functionalism' became a major explanatory principle in the social sciences for a certain period in the first half of the twentieth century. Since then, however, it has fallen into such disfavour that the very word is almost taboo.

Nevertheless, this way of thinking does allow for differences between the public rationale for an institution and its actual effect in practice. The danger is, of course, that the enthusiastic social anthropologist is tempted into inferring concealed positive 'functions' for institutions or customs that are blatantly 'dysfunctional' in their outward form — inter-state warfare, for example, or celibacy of the clergy. But in the present case I am not looking for hidden social forces beneath the surface of ordinary life.

In effect, what I am applying is just a generalised version of the Law of Unintended Consequences. As everybody knows, a research project typically produces knowledge, technical capabilities and personal expertise that can serve a number of other purposes besides the one for which it was undertaken. At any one time, numerous such projects are being carried out by diverse knowledge-producing organisations in a great many different fields of science and scholarship. The incidental 'by-products' and 'spin-offs' of these projects aggregate into a steady stream of knowledge, technique, personnel, etc. These come to be relied on for the regular performance of various other, unrelated social practices. So by supplying these resources science actually performs a variety of 'functions' which it was never consciously purposed to do. In more metaphorical language, it plays a variety of 'roles' which it was not allotted in the original script.

Of course, some of these unintended roles may well be considered 'instrumental', in the sense that they are actively enlisted in the service of particular interests. Thus, for example, computational techniques developed in high energy particle physics are taken over and applied in the design of weapon systems — and *vice versa*. Conversely, some of the products of instrumental science, such as social survey methods developed for use in the marketing of consumer goods, may be employed in very 'pure' research —

say, into the sociology of religion. As I have already pointed out, it is not feasible nowadays to construct an unambivalent classification of scientific entities — items of knowledge, techniques, organisations, people — along the traditional 'pure science — applied research — technological development' axis. Indeed, it is a part of my thesis in this book that all these old categories are becoming hopelessly mixed.

To this we may add the purposes of education because, as we saw in the last section, it is proving so important in our multicultural society. It is the function of school and college education that it should pass on to the next generation the knowledge foundations upon which the nation's culture could best be built. Then they may be able to support and encourage appropriate ways of learning science. Of course we aim to apply this in a way that the immigration of new citizens, as well as our more settled communities, will see that science too, although full of ambiguities, plays an important part in our civil society and also of our changing culture. (This will be discussed in more detail in several other sections.)

The key point, however, is that some of the functions/ roles filled by knowledge production in modern society are not 'instrumental' or 'pre-instrumental' in any ordinary utilitarian sense. But they certainly are not 'useless', because society itself could not operate satisfactorily without them. The fact is that these uses are collective, and not separable from their culture or context. Although they benefit the community as a whole, they do not add value to any particular societal institution — not even the nation-state. A careful analysis may reveal the part they actually play in the polity, the economy, or the general cultural arena, but not as a recipe, a blueprint, a script or an agenda for the principles of how this role should be filled in principle.

Is this just an intellectual fantasy, or an ideological construct? Have we got in mind some naïve or hopelessly outdated model of the social order? Probably, most people do. Well then, let us look at some of the things done for us by science — and remember I mean here by science an *'organised knowledge produced systematically by socially accredited groups'* — other than making some of us healthier, richer and more powerful than any of us were before.

Scientific world pictures

In the first place, science enriches society with influential, trustworthy and new generalised knowledge. Scientists complain about the lack of public understanding of their work, and yet many people have learnt to see the world through their eyes. We immediately think of the revolutions of thought initiated by Copernicus and Darwin. But these were only the most striking cases of a historical process that occurs all the time, on every scale. Note, for example, how nineteenth-century biblical scholarship transformed public perspectives on revealed religion, or how twentieth-century economic theory (right or wrong!) has rewritten the language in which even the tabloid newspapers pontificate about national budgets and international trade.

In the short run, people are often indifferent or actively hostile to new scientific 'world pictures'. In the long run, however, these are profoundly influential. Over a period of years, novel images, concepts, modes of thought and significant facts seep almost unnoticed out of science into the public's ways of expressing themselves. In due course, we wake up to the realisation that the supposed nature of things has changed significantly. Thus, most literate people nowadays have entirely different conceptions of human origins, conditions and capabilities from what they would have had a couple of centuries ago. Some of this new understanding has quite ancient roots. But it also includes images such as the DNA double helix, or the brain as a computer, that have only been visible to science for two generations. And even those who publicly denounce the findings of evolutionary biology, genetics, psychology, anthropology and sociology demonstrate by their vehemence just how seriously they take them. In sum, they immensely influence the way that we locate ourselves in nature and on our planet.

It is not that the latest world pictures painted by science are held to be more realistic, and thus rightly more influential, than the images they superseded. That is a truism. Of course what one now believes seems more credible than what one previously believed without even noticing that one's mind has been changed. Credibility is normally judged according to instrumental criteria. Almost all the philosophical debate about whether science is 'truly' to be believed is narrowly centred on the natural sciences and their associated technologies. There, almost by definition, practical reliability is the acid test.[3] Strong realists typically refute their opponents by asking rhetorically whether they would feel safer in a plane designed and flown by sincere anti-realists — and set off for the airport without even waiting for an answer. When they and we get back, in a later chapter, we can discuss this matter again and more seriously.

Even in its narrow meaning, science is never perfectly certain. Defined more broadly, it can be very insecure indeed. Even well-organised knowledge produced collectively by socially acknowledged experts is not necessarily sound. Consider, for example, the official teachings of a college of theologians, the national glories chronicled by a school of historians, or the aesthetic principles espoused unanimously (for a period) by the current crowd of art critics. Each of these is a knowledge system with the formal attributes of a 'science'. Within its own limits it may well be logically coherent and based on known facts. It may present an attractive picture of a certain aspect of things. But worldly wisdom counsels against putting our trust in it, to the exclusion of all others.

It makes no sense to ask for what purpose this sort of knowledge might have been intended. Generally speaking, the instrumental spirit in modern science has little interest in creating 'world pictures' as such, although there is a fine and continuing stream of such books. Technoscience is too fragmented institutionally and intellectually to unite in

---

[3]   Ziman, J. M. 1978 *Reliable Knowledge: An Exploration of the Grounds for Belief in Science*. Cambridge: Cambridge UP.

such pursuits. Past autocracies have endeavoured to redraw the universe scientifically in their favour. In the event, however, they have failed to suppress the incidental emergence of new ideas which were subversive to one or other of the coeval ideas. Even so these upsetting ranges of ideas were not always produced by would-be political revolutionaries. The prophets of technocracy — H. G. Wells for example — did not even attempt to produce the scientific and technological knowledge whose social applications they urged so vigorously.

Of course many great figures in the human sciences, from Plato through Hobbes and Rousseau to Marx, fancied themselves as societal engineers.[4] But the lasting intellectual influence of their ideas should not be confused with the hapless political blueprints they originally aimed to justify. I am sceptical of general sociological theories that take the first historical signs of the appearance of a new body of scientific knowledge as irrefutable evidence of the instrumental influence of otherwise hidden social forces whose later emergence would seem to be facilitated by that knowledge.

Nevertheless, our pluralistic culture is enriched and strengthened by the diversity of the scientific images that crowd its screens. Our protean selves[5] thrive in a many-splendoured environment where a hundred flowers may bloom. But we crave for more substantial fare than the innumerable fancy snacks generated by a crowd of scholars, inventors and artists, each of whom is intent on doing their own 'thing'. We gain greater nourishment from the products of more coordinated intellectual labour. The various world pictures produced by 'science' are often misleading, or mutually contradictory. But they do present us with a choice of conceivable 'worlds' within which to realise our many-sided personal identities.

Let me give an example. In 1979, James Lovelock pointed out that all the components of the Earth, living and non-

[4]   Popper, K. 1945 *The Open Society and its Enemies*. London: Routledge and Kegan Paul.
[5]   Lifton, R. J. 1993 *The Protean Self: Human Resilience in an Age of Fragmentation*. Chicago IL: University of Chicago Press.

living, interacted so strongly that they should be considered co-evolving constituents of a single system.[6] In a certain sense, it was as if the whole planet were a living organism, capable of maintaining a high degree of dynamical stability over billions of years. To encapsulate this idea (and following sound astronomical tradition), he proposed that it should be given the name of an ancient Greek goddess. This was *Gaia* — the Goddess, Mother Earth.

The Gaia idea brings together knowledge gleaned from numerous diverse sources. It claims to coordinate findings from geology, geophysics, palaeontology, ecology, atmospheric chemistry, marine biology and numerous other highly respectable natural sciences. So it is a genuine scientific theory in the narrow meaning of the term. It is true that many of the scientists specialising in these fields have been scornful of the notion that Gaia is 'alive'. But this is essentially a semantic issue, on which nothing of real significance depends.

What Gaia clearly does, is to provide us with a novel and extraordinarily evocative 'world picture'. It gives substance to our intangible desire for a holistic conception of nature, which would include our human selves. Apart from the universe itself, it is the biggest entity in our experience. On the one hand, it provides an outer frame for all those thoughts about our environment on which so many people would now like to see strict social action. On the other hand, it is sobering to our inmost conceptions of ourselves to realise that we are mere parasites on the great globe itself.

Could there be a scientific idea with richer potential for social influence?[7] Indeed, the Gaia idea seemed to some people so worthy of respect and consideration that they began treating it as 'sacred'.[8] In effect, it had begun to perform some of the cultural functions of a *religion*. But this is just one example of the tendency for bodies of knowledge to

[6]   Lovelock, J. E. 1979 *Gaia: A new look at life on the Earth*. Oxford: Oxford UP.
[7]   Midgley, M. 2001 *Gaia: The next big idea*. London: Demos.
[8]   Primavesi, A. 2000 *Sacred Gaia: Holistic theology and earth system science*. London: Routledge.

spread beyond their established areas of competence, and to assume social roles for which they were not originally equipped. Thus 'creation science' — let us not quibble about its right to this title — tries to extend beyond its legitimate theological base into the domain of evolutionary biology. Conversely, some evolutionary biologists claim quasi-religious authority over moral issues which lie far outside the boundaries of their scientific expertise.[9] Society evidently has a place for such pluralist roles, even though they are played impromptu by actors who were never scripted to perform them.

Sometimes, of course, all that science has to offer is a circus of wonders, a cabinet of curiosities as they used to be called in the early days of museums. The restless mentality of our civilization seems to require a regular diet of amazing discoveries to match its insatiable appetite for novel inventions. The past two centuries have abundantly fulfilled both needs. New developments in cosmology, particle physics, plate tectonics, animal behaviour, cognitive science, etc. have more than equalled progress in medicine, engineering, information technology, agriculture, military technology, and so on.

The BBC decided, in 2004, to axe *Tomorrow's World*, a long-running TV programme that showed both forms of new knowledge as if they were the two sides of the same coin. The tendency, of course, is to stress the instrumental potential of all science, however far from any conceivable practical application. But this 'pre-instrumental' discourse is only the transparent patter of would-be salesmanship. And it is not really necessary. Our pluralistic society is not just a commercial marketplace. People are not just being conned into supporting very expensive pure science because it might one day cure cancer or fly us as tourists to Mars. Nor are they being softened up mentally to buy into the very latest gismos being advertised on the internet. But they do value and celebrate emergent novelty for its own

[9]   Midgley, M. 1992 *Science as Salvation*. London: Routledge.

sake, in the natural and human sciences as well as in the creative arts. That is a very broad sweep.

As a consequence, 'the search for truth' has acquired a romantic aura comparable to a religious vocation. It applies right across the knowledge spectrum, from highbrow physics to highbrow cultural studies. Many young people enter science intent on making discoveries in the spirit of missionaries intent on saving souls. They do not think of themselves as being engaged in 'the production of knowledge', nor as serving the practical needs of humanity by their contributions to technoscientific industry. That more realistic assessment of their place in society only comes later. Naturally, they want to succeed professionally, but only a few start off with the idea that their calling might be a way to riches or personal celebrity. The pursuit of truth only morphs into the pursuit of a Nobel prize when the latter suddenly emerges as an attainable token of the former.

Wonder is the word for this non-instrumental product. It may be a genuine cultural artefact, even if it can't be given a stock-exchange rating. Utilitarianism, of course, has no place for such frivolities at all. Yet we all know, in our hearts, that these are intangible goods that give us as much sustenance as food and drink, and our forebears six centuries ago, in the same mode of wonder about the world around us, called it 'Anima Mundi'. We have become so accustomed to such ideas that our existence would seem insufferably dull without them. What follies would our children be committing if they were not so fixated on dinosaurs? Populist politicians sometimes denounce the far-out findings of pure science — the strange particles, the quasars, the pre-Cambrian fossils, the traces of buried villages, the dusty old charters, the reconstituted texts of eminently forgettable philosophers — as cultural luxuries that can only be appreciated by the *cognoscenti*. In the long run, as we have seen, these findings are domesticated and incorporated into other-world images that are sometimes shared widely enough to become embedded in the mass consciousness.

The real worry is that this drama of continuous disclosures, like a dance of seven veils, cannot go on for ever. Like economic growth, it must one day come to a stop. It was a brave, perhaps foolhardy soul who suggested recently that the End of Science was nigh.[10] Well, he was just a journalist, and not intimidated by the fact that this prophecy had been uttered by scientists themselves at regular intervals for centuries — and confounded each time by further spectacular discoveries or inventions. All I would say is that the most effective way of making it happen would be to insist that all knowledge producers must henceforth articulate their research plans, and state their instrumental objectives, before they set out on any of their voyages into the unknown!

Sensing danger

One of the prevailing features of modern society is anxiety about impending disasters. Scientific knowledge plays a major role, both in sensing unfamiliar dangers and shielding us against the worst of their outcomes. Of course, human life has always been a risky business, but we have plunged ahead regardless. Nowadays, so it is claimed, we live in a 'risk society',[11] where the perception, assessment and avoidance of 'unacceptable' risks[12] has become one of the major functions of science.

Quite obviously, public concerns about health, energy supplies, food resources, employment, nature conservation, and so on are the driving forces for an immense amount of directly instrumental science. Many hazards are, so to speak, routine. The insurance industry thrives on them. So society invests in sciences such as epidemiology, meteorology, seismology and economics in order to predict and limit their effects. A modern government would be considered criminally irresponsible if it were not actively fostering the production of knowledge relevant to all such

[10]   Horgan, J. 1996 *The End of Science: Facing the limits of knowledge in the twilight of the scientific age.* New York, NY: Addison Wesley.
[11]   Beck, U. 1986 (1992) *Risk Society: Towards a New Modernity.* London: Sage.
[12]   Ravetz, J. R. 1977 *The Acceptability of Risks.* London: Council for Science and Society.

concerns. The same applies to non-governmental and commercial bodies. The safety of consumers, clients, patients, travellers, experimental subjects and other 'customers' is one of the foundation principles of technoscience. From the basic research laboratory to the post-release market analysis, this is an immutable law of knowledge production.

Indeed, the necessary precautionary principle and regulatory procedures based upon it, are themselves increasingly active topics for systematic scientific work. Risky situations are where 'science, technology and society' typically interact, often explosively. Such interactions now form the subject matter of a significant proportion of the 'STS' literature — that is, in the journals and conference proceedings devoted to 'Science and Technology Studies'.

At first sight, one might suppose that this is where instrumental science rules unchallenged. But, to understand how things might go or have gone wrong, and especially to allocate blame, we must draw upon a wide range of other scientific disciplines. So non-instrumental knowledge, from fields such as law, moral philosophy, religious ethics, social psychology and cultural anthropology, play a significant part in interpreting the human factors in risk situations. Indeed, 'risk' is a sure-fire catalyst for such social conflict. As we shall see, non-instrumental science is the principal source of the non-partisan technical expertise often required to resolve such conflict.

But non-instrumental science also plays another major role in the calculus of danger. The simple fact is that the *initial* awareness of a *possible* hazard cannot be programmed instrumentally. How can one set out to become conscious of, and assess the likelihood of, a misfortune that one does not even expect? Generalised early-warning devices are notoriously unreliable. In fairy tales, even magicians typically disregard the unexpectedly disastrous scenarios which they can see in their crystal balls!

In practice, alas, an actual disaster is usually the first sign that we have run into danger. Only then, far too late, do we realise how easily it could have been foretold. Nevertheless, there are a few situations, some of them of great importance, where a serious hazard has been perceived, and guarded against, before it strikes. And such intimations often come from quite unexpected places.

For example, who could have even thought about climate change before Svante Arrhenius, an almost perfectly academic Swedish professor, began to calculate the effects of the absorption of solar radiation by gas molecules in the atmosphere. His research was motivated by curiosity about the causes of the Ice Ages. In no sense was it directed towards making a 'useful' discovery, or interpreting it in such a protective spirit. That was a century ago. But although the general idea of a 'greenhouse effect' soon became familiar amongst physicists, and was applied mathematically by meteorologists, its broader significance was largely ignored. It was not until the 1970s that the possibility of significant global warming became a controversial scientific issue in relation to energy supplies.[13] And even then, the study of ancient climates using their traces in the geological record left behind in ice-cores and fossils, was still regarded as purely 'academic'.

Nowadays, of course, many thousands of scientists, all over the world, are busily investigating every aspect of climate change in a thoroughly utilitarian mode. In effect, as James Lovelock now puts it, Gaia is now the patient.[14] The very title of the *Intergovernmental* Panel on Climate Change show that this instrumental enterprise is now high on the political agenda. Science not only has its own place in society; it is in a red hot seat!

This is technoscientific risk assessment on the grand scale. We need not comment on its technical strengths and weaknesses. Nor need we go into the enormously contentious political questions that now surround it. We shall discuss later the challenge of establishing the trustworthiness of expert informants on just such controversial matters. The key point here is that the initial perception of such

[13] Ziman, J. 1979 *Deciding about Energy Policy*. London: Council for Science and Society.
[14] Lovelock, J. 1991 *Gaia: The Practical Science of Planetary Medicine*. London: Gaia Books.

long-term hazards is a typical non-instrumental role of science. The same may be said of many other environmental dangers, such as the decline of biodiversity. How would these have become apparent — and now the active concern of every dutiful school child — without the collective labours of thousands of naturalists, taxonomists, ecologists and other scientific observers, most of whom were working to no more 'useful' purpose than a better understanding of the natural world.

The world pictures produced by science are seldom static. They extend back into history and can be projected into the future. They offer plausible scenarios of how things might be, within a few years or a few centuries. And some of those scenarios are grim indeed. So we are alerted to dangers that are already germinating in the present, and occasionally given hints on how they might be avoided in the future.

This, surely, is one of the traditional functions of the human sciences. As an instrument of the official authorities, they typically serve the established social order. But even the most carefully contrived images of normality can be disruptive. For example, some contemporary social scientists try to portray free market capitalism as the best of all possible worlds. But the striking contrast between their cheerful pictures and the actual lives of most people is a provocation to revolutionary change. The stance of scientific objectivity is soon discredited by the compromises they have made with reality. In effect, their blatant instrumentalism becomes a potent weapon against them.

In a genuinely pluralist society, where there is no ruling ideology, the human sciences perform a direct non-instrumental role. Through their alternative world pictures they convey a critique of the status quo and of the various actors that contend for power. History, for example, warns our rulers of the likely consequences of their present policies. Economic theory is not a reliable guide to small scale practice, but reveals some of the risks and uncertainties of current doctrines. The utopias and dystopias painted by political philosophers are only hypothetical, but they present vivid images of genuine hopes and fears. Sociology and anthropology often report on paradises that are being lost, and are not likely ever to be regained.

It might be claimed, of course, that when science is openly critical of the established social order it is simply operating as the tool of the revolutionary forces of the day. This is often how criticism is countered by the defending class or political party. It is interpreted in an instrumental spirit, where 'They would say that, wouldn't they?' passes for a fatal riposte.

Paradoxically, vulgar Marxism holds to the same principle, i.e. that all knowledge is produced to serve the material interests of one or other of the groups struggling for control of society. Taken literally, of course, this would undercut its own claim to be speaking truth to power. Indeed, if accepted absolutely this principle would also undermine the whole political philosophy of the pluralist society. So although it's a serious point, to which we shall necessarily return, it requires somewhat stronger backing than mere assertion. Certainly we shouldn't assume at this stage that it completely invalidates my argument that scientific knowledge, generally defined, is often produced in a non-instrumental mode that gives it important roles in society.

Scientific attitudes — sometimes scientistic

Another of the non-instrumental functions of science is to inject 'scientific attitudes' into public disputation. Here I don't mean the actual substance of scientific knowledge. I am not referring, for example, to the way that biomedical knowledge and techniques have completely changed the terms of reference of all political controversy about the ethics of contraception. Nor need I run through all the sciences implicated in the global debate about Climate Change — not just the physical and earth sciences but agriculture, economics, law and political theory. Scientific knowledge, in the broadest sense, is woven into the fabric of public life. Some of the speeches in the British House of Lords sound as if they were meant for the Royal Society — a body of which many life peers are also Fellows.

What is largely overlooked is that the way in which issues are publicly discussed and disputed owes much to science. It would not be fair to the great debating forums of the past, such as the Classical Greek agora or the Councils of the Early Church, to suggest that they were deficient in the presentation of well-reasoned argument. After all, it was the Greek sophists and their scholastic successors who put 'rhetoric' — the science of persuasion — on the academic curriculum. But each era has its own rhetorical style. What counts as a convincing argument today has to satisfy certain general criteria of which the contestants are usually quite unaware. But these criteria vary subtly from one era to another.

Living in a scientific age, we accept the scientific style as completely natural. Indeed, we insist that only an argument that conforms to this style should be considered completely 'rational'. Well, as I have remarked, this only means that it can be clearly articulated and communicated unambiguously to other members of ones community. The important thing then is how consistent this style is with the framework of shared assumptions within which it is supposedly located. But this framework is necessarily tacit. It could only be defined in terms of a more general framework. And to define that framework? Well we need to climb up to a further stage, and so on without end.[15]

In other words, we learn to 'think scientifically' by living, working, and talking with each other inside scientific communities. The practice of modern science has taught people to argue in what we have come to call the scientific style. What is so interesting is the way that this style has spread out of the world of professional research into the much larger domain of public affairs. Science is in the chair of the panel of judges of rational discourse. This is its most influential place in society. It is one of science's major roles.

[15]   Ziman, J. M. 2000 *Real Science: What it is and what it means*. Cambridge: Cambridge UP.

Note, nevertheless, that this role is basically non-instrumental. Scientific knowledge is produced for its substance, not for its style. The goal of research is almost always to answer questions and to solve problems, not just to show how to conduct a convincing argument. That is the peculiar specialty of pure mathematicians and logicians. Sometimes, of course, a scholarly discipline finds itself in a blind alley and gets itself into a tizz about its methodology. But the results of such navel-gazing are seldom applicable elsewhere. Many scholars, for example, regard it as a disaster that the post-modern notion of 'deconstruction' as a general tool of thought managed to spread out of the field of literary studies into all the human sciences.

Of course there are people who hold that the scientific style of argument is uniquely efficacious. They believe that the 'scientific method' is a universal solvent for all problems. Science, they assert, is not just a remarkably powerful instrument of human will: it is an all-powerful means for achieving any goal, however remote or bizarre. Its social role, then, is to take control of society, and try to make it scientifically perfect — whatever that might entail.

As we have seen, this totally instrumental programme is the political agenda of *technocracy*. Its ideological base is *scientism* — the doctrine that the only valid, reliable, credible, true (etc.) form of knowledge is scientific. Typically, this is further narrowed down to exclude most of the human sciences, whose findings seem to scientists so tiresomely vague and/or contradictory. Well, it is a comforting thought that society has, ready at hand, a cure-all for its manifold ills. As we shall see, the famous 'method' of science is an elusive concept, and certainly not wide, deep, or firm enough to carry such a vast superstructure of action. What is more, a tool should only be valued as a means to desirable ends. Neither for people in particular nor for society in general are these ends 'technical', even in a narrowly scientistic sense.

So the wholesale adoption of a self-conscious 'scientific attitude' is not a remedy for an ailing society.[16] It has its good points, but not as an infallible recipe for doing the right thing. Indeed, it sometimes parodies itself. But this is not the place to deplore the vulgar tendency to 'scientise' the presentation of every aspect of ordinary life, from the health-giving virtues of breakfast foods to the wealth-giving sizes of television audiences. Nor need we pause to smile at the pedantic obfuscation of relatively straightforward problems in clouds of elaborate terms, recondite theoretical concepts, or difficult and pretentious mathematical formulae. These are only more superficial signs of the sometimes interesting influence of science on the *mentalité* of our pluralistic culture.

In particular, *instrumentalism* itself has become the predominant principle of life-world action.[17] To behave rationally, so we all now believe, is to choose the right *tools* for the job. The *means* we employ must be well matched to the *ends* they are designed to achieve. They must be based on well-founded knowledge of the facts and causally linked to their desired *consequences*. In this book I continually speak in terms of 'roles' and 'functions'. These are typical of the instrumental metaphors that we use so unselfconsciously to explicate the dynamic of social activity.

The point here is not just that science itself has become a major instrument of policy—that the production of scientific knowledge is so frequently undertaken as a means to a specific end. It is that the very form of this knowledge embodies and exemplifies the instrumental spirit. After all, a large proportion of modern science is devoted to improving our understanding of the general principles of practical technologies such as engineering and medicine. Instrumentalism is also implicated in the seventeenth century conception of the universe as a machine, an idea pick up from the newly invented mechanical clock of the

[16] Grinnell, F. 1992 *The Scientific Attitude*. New York & London: Guilford Press.
[17] Ezrahi, Y. 1990 *The Descent of Icarus: Science and the Transformation of Contemporary Democracy*. Cambridge MA: Harvard UP.

fourteenth century together with its mechanised figures. This metaphor has since been extended to living organisms, to the human brain, and to society itself. How things work is inseparable from what they can do; even better, what we can do with them. And as I have argued elsewhere, much highbrow theoretical research is directed towards improving the 'reliability' of knowledge, as if it might one day have to be put to the test of practical use.[18]

So when we ask for public figures to justify actions, we seldom expect a genuinely 'scientific' answer. But we do expect the scientists to show that what they have done was founded on demonstrable 'facts' — preferably quantitative — articulated in terms of well-established general principles and produced measurable results. In other words, their justificatory reasoning must be structured instrumentally. Religious, ethical and aesthetic doctrines, historical traditions, legal precedents and personal whims are still heavy hitters in political discourse. But they are often dismissed as 'irrational' by comparison with hardheaded practical logic modelled roughly on the scientific style of argumentation.

It is a commonplace that science thrives best in an open society. What is seldom recognised is that political pluralism owes a great deal to the science it fosters. As Yaron Ezrahi has shown in detail,[19] scientific instrumentalism is the deep ideology of modern liberal democracy. One of the roles of science, albeit unintended, is to exemplify this ideology in action. Science actively produces, and puts into practice, immense quantities of reliable, efficacious knowledge in a variety of spheres of life. It thus demonstrates the possibility of producing and putting into practice this type of knowledge in the political sphere, where it is most in demand.

Paradoxically though, instrumental science as such is a dangerous model for scientific instrumentalism. Of course it can be enormously heartening to see a major industrial

[18] Ziman, J. M. 1978 *Reliable Knowledge: An Exploration of the Grounds for Belief in Science.* Cambridge: Cambridge UP.

[19] Ezrahi, Y. 1990 *The Descent of Icarus: Science and the Transformation of Contemporary Democracy.* Cambridge MA: Harvard UP.

research and development project through to material success. Surely, if we can manufacture miracle drugs, super-fast computers and nuclear submarines then we can make similar progress against crime, poverty and political violence. It must be just a matter of assembling a critical mass of scientists, and putting them to work. Indeed, the requisite theoretical knowledge may already be in the archives.

But that way lies *technocracy*! Dazzled by what we think *can* be done, we are easily led to believe that it is what *ought* to be done. That might be acceptable if those with the authority to do such things were always wise and benevolent. But that requires equally thoughtful, persistent 'scientific' consideration of the purposes to be achieved and the 'side effects' of achieving them. OK, so we can build the biggest dam in the world and calculated its profitability according to the latest economic models. But have we adequately evaluated the miseries of personal displacement, the destruction of sacred places, the damage to a multifarious biological environment? None of these can be costed financially, so should we really leave them out of the equation?

The mindset of scientific instrumentalism is not receptive to human values. What is more, instrumental science is blind, by its very definition, to the purposes for which it is to be used. It is focussed entirely on the means for achieving those purposes. As we shall see, this puts it entirely in the hands of any person or persons with the money to buy its services. In principle and in practice, it performs the role of a highly intelligent, extremely expert, but always willing slave. Its duty is simply to serve the interests of its master. All questions of right or wrong, equity or exploitation, kindness or cruelty, wisdom or folly, are for others to decide.

For some aspects of social life this servile attitude may possibly be appropriate. People like that are always needed to get the ordinary work of the world done — and even some of its extraordinary work. In fact, we may all need to be a bit like that for a part of the time. But the important social function of science is surely *not* to teach us 'not to reason why' and to be eternally 'on tap'. And that, surely, is not what a pluralistic society requires of its citizens. This is where non-instrumental science exemplifies other social virtues that are equally vital.

Take, for example, the role of 'theories'. Much science is devoted to the production of general 'maps' of various aspects of the world. These represent our knowledge of the nature and behaviour of things. From them we can read off where we are and what will happen if we take such and such a route. Even when manufactured without that purpose in mind the theories embody the instrumental capabilities of science in the most compelling and compact form.

Now it can rightly be said that scientific theories take the subjective element out of social action. They are abstract and impersonal. So the public official or industrial manager who appeals to them as justification for her actions can claim to be genuinely impartial too. The rule of science is like the rule of law. It replaces the arbitrary irresponsible human factors in the social machinery with perfectly reliable working parts. In a pluralistic society it takes the heat out of many cases of potential friction. And when political actions are challenged, scientific theories (like established legal principles) provide the actors with a framework of public accountability, within which they can show that they behaved responsibly in the circumstances of that day and incorporate it in the book of legal presidents.

Yes, indeed: scientific theories flourish in a liberal democracy—a civil society as we shall see. But they also make feasible the bureaucratic utopia of an orderly society where freedom is constrained only by necessity (as the old Marxists used to proclaim). In reality, of course, this would be a totalitarian nightmare.

Fortunately, science provides the antidote for its own poison. In its non-instrumental mode it is as much concerned with the *deconstruction* of theories as with their construction. Familiarity with science can be a sobering experience. We learn how difficult it is to get things right, how little we know for sure, and how uncertain our theories really are. We are reminded that most generalisations are mere dogmas, to be doubted and tested, and then doubted again — unto destruction!

This scepticism applies even to supposed 'facts'. In the courts of public accountability, these are obviously vital. But only if they are fully disclosed and subject to systematic further investigation. Indeed, as Ezrahi points out, 'transparency' is another of the scientific attitudes that contribute to the viability of liberal democracy. 'Public knowledge', as I labelled science in an early book, is what holds science together as a social institution.[20] This is not just what many people happen to know: it is what they could and should get to know to judge the credibility of a factual or theoretical claim in situations where damage is at risk. The same applies to all that is asserted by public figures to justify their actions. It is not sufficient, as so often in technoscience, for them to argue solely in terms of actual outcomes. *'It came out all right on the night, so what are you worrying about?'* In a fully accountable political system, as in non-instrumental science, every step in the decision process has to be open to public scrutiny.

Perhaps the most difficult lesson that a pluralistic society has to learn from science is how to deal with dissent. The production of truly reliable knowledge not only requires a sceptical attitude towards 'the received wisdom': it also requires a rich and varied selection of possible alternatives. In some ways, the creation of testable hypotheses is the final name of the game. Much of the delight of research is having shiny new beautiful thoughts. *Originality* is its most-prized, but most elusive norm.

So every genuine scientific community — and remember that I include all fields of collective scholarly endeavour — is in a permanent state of simmering discontent. Its institutional duty is to be disputatious. One has to learn to listen patiently to wild conjectures and entertain them seriously, even though one thinks one knows all the excellent reasons for dismissing them outright. One even has to try to unthink

[20] Ziman, J. M. 1968 *Public Knowledge: The Social Dimension of Science.* Cambridge: Cambridge UP.

ones own beautiful thoughts, yet maintain them against apparently sound objections. One has to publicly applaud the pronouncements of the established authorities without holding back from publicly pricking the balloons of their intellectual complacency. One has to engage vigorously with ones rivals on the scholarly plane, whilst rejecting every temptation to make personal jibes. One must continually affect a public scholarly demeanour of modesty and humility, even though in private one is driven by pride and ambition. And so on.

What is usually less obvious, however, is that this competitive individualism is always balanced by cooperative communalism. One has to think of science as a joint enterprise. However attached one may be to ones findings, they belong to the whole community. Once it is verified, valuable knowledge must be shared with others, even professional rivals. If two heads think better than one, then let them work together. We are always standing on the shoulders of others — many of whom are our contemporaries — so we strengthen ourselves by strengthening them. Pride and confidence in what we do together is more appropriate than personal vanity. It is required to be fragile to some extent, and also to be shared without fear.

These are hard acts to sustain. Few scientists really achieve such moral stature. But these are attitudes that are also of great value to the civil society and in public life. In particular, they are just what is needed to combat instrumental scientism. One of the most important social functions of non-instrumental science is to counter technocratic arrogance with expert criticism, well-founded scepticism and imaginative alternative scenarios. In the language of parliamentary democracy, it has to perform the role of a 'loyal opposition'. As we shall see, this dissident function also requires appropriate institutional arrangements. But the will and capacity of individuals to support it comes from the attitudes they acquire during the education in, and practice of science which they then transmit to society at large .

The fact is that the production of scientific knowledge is itself a pluralistic enterprise. At a later stage we will consider the notion that it might even provide a model for society at large. Anyway, to do that, it would need to take off its technoscientific blinkers, and look much more widely. The laudable scientific attempt to reduce the universe to universals cannot hope to succeed. But as a participant in its failure one learns in the hard way just how valuable those obdurate attitudes that we call 'human values' really are. In other words, contrary to what most scientists believe, one of the strictly non-instrumental social roles of science is to affirm and reinforce the moral and epistemic pluralism that underlies our civilization.

FOUR

Producing Knowledgeable People

Training scientific experts

The production of knowledge is inseparable from the production of knowers. To make new science one has first to learn old science. Old science can only be handed on by those who already own it. To own knowledge, one has to have had a hand in the making of it. So education and training, teaching and learning, scholarship and research, are all wrapped up in each other.

From the dawn of written history, the official function of scholarly institutions has been to train expert practitioners. Through their schools, passed the doctors, priests, lawyers, clerks, and school teachers, needed to run a civilized society. To facilitate their instruction, knowledge had to be formalised. Soon it was collectivized, and in due course, the mastery and furtherance of this science was acknowledged as one of the major functions of the institution of science.

That may seem like only another *Just So* story. But it does broach the chicken/egg question: which came first, teaching or research? Let me approach that from a pragmatic perspective, where the two activities appear to be so intertwined that they have obliterated the traces of their historical sequence. The simple fact is that the great majority of scientific knowledge is produced in universities and other institutions of higher education. In many cases, this knowledge is apparently quite 'useless' and irrelevant to lifeworld practice. Much of it is only vaguely related to what is being taught to the majority of the students. Nevertheless, this mode of scientific activity helps significantly in the preparation of skilled practitioners and specialised experts for the influential positions that they will eventually occupy in the continually changing knowledge-based social order. This seemingly incidental association of science with professional training is actually one of its major non-instrumental social roles.

Needless to say, the Juggernaut of instrumental technoscience has an insatiable appetite for highly specialised technical workers. These certainly need to be familiar with the latest theories and trained in the most up-to-date techniques before they join their research teams. But what about the much larger numbers of ordinary practitioners in skilled professions such as engineering, medicine or law? Much (though not all) of the knowledge that they apply every day certainly derives from high science. But they are seldom directly challenged by researchable problems, and are not expected personally to actually produce new knowledge — even if they had the time or the inclination to do so. What then do they gain from having been trained in an atmosphere of active discovery and invention, by teachers who are heavily engaged in the demanding and absorbing labour of research?

The accepted answer to this question is: '*a very great deal*'. At least that is the unanimous response of leading scientists, scholars and practitioners, boomed out through the audio systems at innumerable celebratory events. Well, they would say that wouldn't they, given the close institutional and career links between academia and the learned professions? As it happens, I entirely accept this answer myself, although it's not so easy to explain why.

The conventional line of argument would seem to be as follows: as a traditional craft becomes more 'scientific', it becomes subject to the norm of 'originality'. In the struggle for personal recognition, the scholars and researchers put a high value on novelty. For better or worse, continuous change — 'progress' — becomes the normal state of the art. Practitioners owe it to their patients, clients and customers to be up-to-date. So they need to be trained by people whose active participation in research makes them familiar with all the latest theories and techniques.

That is all very well, but it can only work for a time. The front line of scientific and technological progress moves too rapidly. Within a few years, even the best-trained practitioners have been left far behind! Then there is a need for refresher courses, in-service retraining and other regular facilities for keeping skilled technical workers systematically informed about the science of their craft. So runs the conventional rationale for combining practitioner training with research, which turns out to be of limited scope.

A more convincing answer might be that society requires more of its skilled practitioners than mere technical expertise. It requires at least a small grain of *enlightenment*. I don't just mean human understanding of human problems, or even managerial understanding of managerial problems. I mean an appreciation of the true strengths and weaknesses of their specialised capabilities.

People practising any science-based calling need to remain aware throughout their careers of the gaps and uncertainties in their knowledge base. They need to be actively receptive to new developments and continuously re-educating themselves so as to learn about them as they emerge. Ideally, they should always feel that by making use of new knowledge — or by choosing intelligently not to use it — they are helping to produce it. They should not be expected to approach their regular duties as if they were themselves research scientists, brimming over with insatiable curiosity. But they do need some of that attitude of mind, and that is the most valuable non-instrumental outcome of professional training in the atmosphere of scientific openness and change that are characteristic of institutions engaged in the active production of scientific knowledge.

Who really knows about it?

Our so-called 'information society', however, is not a smooth-running machine. As we have seen, it is a minefield of risks and a jungle of conflicting interests. To deal with these it needs more than skilled technoscientists and enlightened practitioners: it needs independent *informants*. By this I mean people who can be trusted to provide highly specialised information on complex matters of dispute or concern. In other words, our pluralistic culture relies, at crucial points, on the testimony of *experts*.

What is more, the law-governed practices of liberal democracy presume that *non-partisan* experts can always be found to fill certain public roles. There is a continual demand for expert witnesses, legal assessors and arbitrators who are not attached to either side in a contested case. On occasions, consultants and advisers are called upon for an 'outside' opinion on some controversial but highly technical matter. Time and again, the 'talking heads' on the media are identified as well-informed people who are ready to speak their minds publicly without fear or favour. Of course, even the most knowledgeable experts often disagree with one another, so their views cannot be supposed to be unchallengeable. But the business of the court of law, or commission, or panel, or debating forum, would never be concluded if their professional integrity or sincerity always came into question.

I am not suggesting, of course, that such expertise could possibly be *absolutely* 'objective'. That could only be true of philosophical fantasies, not of lifeworld realities. Scientific knowledge is produced by normal human beings working together in normal social institutions. Their lives are buffeted and shaped by their societal circumstances. And however sincere they are, however hard they try to be completely dispassionate, they are always adrift on the unseen tides of social change. Whatever they say or do, even about matters of tangible fact, is bound to be biassed, one way or another, from their deep regard for the science involved, to their rejection of everyday maxims.

So, some of the elementary sources of possible prejudice can be recognised and systematically countered. Just how far this is achieved is a matter for serious study, not for ideological dismissal or cynical disdain. And much the most obvious factor is material interest. As in a court of law, little weight is attached to the testimony of an informant who has organisational, family or financial connections with a party to the issue. People simply will not believe in the 'objectivity' of a technical expert who is known to be financially or socially involved with one side or the other of a dispute.

Now we come to a point that will be discussed at greater length later. In essence it is very simple. By its very definition, instrumental research is undertaken precisely to serve the interests of those who organise, plan, fund and hope to benefit from it. For that reason, scientists engaged in that mode of research can seldom be trusted to give entirely independent expert advice on controversial issues on which they are surely knowledgeable. This is not a point that can be finessed epistemologically by reference to the 'objectivity' of scientific knowledge. It is a straightforward matter of street-wise credibility and practical prudence.

One of the major non-instrumental roles of science in a pluralistic society is to provide just such non-partisan expertise over a very wide range of issues. This is a very demanding role, since it requires people with a correspondingly wide range of specialised technical knowledge — the sort of knowledge that can only be gained and kept up-to-date by active engagement in scholarship and research. In a society where the production of knowledge is a major industry this would seem relatively easy. But the further condition that this production process should be effectively 'disinterested' is much more difficult to satisfy.

Expertise in risk assessment

To fully appreciate this difficulty, consider again the problem of dealing with risk. It is in the nature of the human condition that hazards cannot be eliminated from it. The best that can be done, in the end, is to arrange things so that relatively uninstructed people can be given sufficient information so that they can decide for themselves whether or not the residual risks are 'acceptable'. For more than twenty years, this has been the buzz word in all high-level discourse on the subject.[1] But who is supposed to be doing this 'accepting'? Not, one would hope, just the people whose actions generate the possible hazards, or who have a duty to 'manage' their consequences for others. In an open, morally ordered society, risks can only be 'accepted' by those most endangered by them.

Very frequently, this acceptance is tacit. If people voluntarily undertake activities that might obviously harm them, then we are entitled to assume that they know what they are doing. But our culture nowadays is so elaborate and interdependent that it is not at all obvious what perils we might be subject to even through our own actions, let alone the actions of others. To impose a known danger cavalierly on other people is clearly unjustifiable.

So even when people have supposedly signified their 'acceptance' of this risk, this must be on the basis of their being well informed of its nature and scope. This applies particularly to unintended damage due to 'human error'. When there is an accident, we would usually ask how it was that the relevant technical system failed to protect us from harm. But there are always people and institutions in the loop.

Needless to say, such concerns consume a vast amount of mental and social energy. Much discussion of the 'place of science in society' is actually about this sort of thing. As we have seen, elaborate analytical and political procedures are now devoted to detecting, analysing, and meliorating the dangers to which we put ourselves in maintaining our present life style.

Indeed, the assessment of risks is not a single province of expertise. The 'acceptability' paradigm assumes an extended social process, in which a great many different people are involved, each exercising a different mode of expertise. Thus, in the first instance, a danger has to be *suspected*. In the case of Climate Change, this took a long time. As we have seen, Arrhenius's argument was very speculative. He realised of course that the concentration of the main

---

[1]   Ravetz, J.R. 1977 *The Acceptability of Risks*. London: Council for Science and Society.

'greenhouse gas', carbon dioxide, was affected by the industrial use of fossil fuels, but calculated that this was not serious at the then rate of coal consumption.

But then a much more intensive expertise in meteorology was required to *perceive* this effect in action. It took a somewhat wider view of historical climate change — a very marginal academic specialty, whose expert practitioners were likely to have no established discipline — to *formulate* the possible consequences of not attending to it. As a result, experts from many different fields — astrophysics, marine biology, forestry, combustion engineering, etc., not to mention the powerful corps of computational model-makers — had to be brought together to *analyse* the system. Then the social and economic dimensions of the risk had to be *assessed* by the corresponding experts, and the results of all these exercises *publicized* by some people who had acquired from experience an uncanny understanding of the grand politics of the situation. And to close the circles or advance the helices of consultation and decision, the various interest groups involved needed experts to acquire broad knowledge of the process as a whole and to *advise* them on the way it was going.

Because of its awesome implications, climate change is probably the most complex risk assessment process that has ever been attempted. But the same applies wherever science makes itself felt in society. A great variety of expertise is required to *suspect, perceive, formulate, analyse, assess, publicize,* and generally *advise on* socially significant risks.

Risks are so very uncertain

The difficulty of talking about expertise is that it always claims to be uniquely authoritative. '*Thus and thus it is*', the expert informs us — and who are we to question her further? But that authority can never apply to risks. If they were perfectly clear cut and unequivocal, they would not be 'risky'! So even the most eminent experts cannot be considered absolutely reliable.

In practice, the situation is always much worse than merely opaque. Every step in a risk assessment, from initial perception to policy action, is open to theoretical or factual debate. That is inevitable, because all risk discourse refers to an unknown future. Its main product is the innate unpredictability of what one would like to foresee. The more seemingly precise one makes ones expert analysis, the more hostages it presents to deconstruction by other experts. In fact, it is their bounden duty to find the holes in ones arguments, and pick away at them in the fond hope that they will fall to pieces. That is what 'being scientific' often really means.[2]

Much of the time, of course, the information that leads people to accept risks is just the received wisdom of the day. Alcohol, so most people believe, is less dangerous than 'hard drugs', and tobacco less dangerous than cannabis. Well, that's what everybody 'knows' to be true, including the many doctors who surely would never drink or smoke if they thought otherwise? It takes a chorus of critical, heterodox expertise to overthrow such a widely held notion.

In any case, risk is not just a 'technical' parameter in the trajectory of our lives. In the crudest arithmetical sense, it is a number you get when you multiply a probability by a measure of value. Having envisaged an adverse event — not at all a trivial mental operation — you must try to work out how likely it is to occur. That is usually the factor on which the experts focus, because it is thought to be just within the bounds of technical reasoning. But the value — typically a large negative quantity — to be assigned to this putative event is much more disputable.

The 'cost' of an accident, disaster or other undesirable circumstance not only involves diverse personal and social preferences, like being more or less healthy, more or less efficacious as a social institution, etc. It also requires an image of the general context in which this unfortunate future event might be supposed to occur — a world of competing, warring, nation states, say, or one that has achieved some degree of global coordination and wealth sharing. It would contradict our expertise as students of humanity to arrive freely at a consensus on what this number ought to be. Even the notion that it could always be expressed in dollars, or euros, rather than say, 'Quality Adjusted Life Years', is wide open to honest and well-founded doubt.

In the real world, then, and in all virtual worlds that can seriously claim to mirror it, expert opinion on risky situations is typically controversial. The supposition that it could ever be perfectly definitive or 'objective' is a mirage, for it is not generated by 'objects'. In its most vital parts, it is produced by human beings with human strengths and weaknesses. Certainly, we experts, in our particular fields, strive mightily to bring to bear on it our most powerful reasoning tools. Whatever the concept of 'being scientific' implies in our context, we endeavour to apply it sincerely. And as we shall see, the social framework in which many experts operate can be a very effective guard against obviously 'subjective' deviations from a generally defensible communal view.

Nevertheless, there are many modes of rational argumentation, and always numerous ambiguities of fact, theory and moral principle. These uncertainties always leave ample space for much genuine difference of opinion, even amongst well-qualified experts. And because what happens in this space inevitable affects the material prospects of numerous individuals, groups and organizations, it is bound to be a field of operation for the commercial or political 'interests' vested in these outcomes. That is where 'society' really messes up our 'science'.

How do we find experts we can trust?

The corruption of expertise by self-serving bias is fully appreciated by the streetwise citizens of our world. Indeed, some of our societal institutions are designed to make it evident and then factor it out of the equation. In regulatory disputes, for example, it is customary, and in no sense illegitimate, for experts to be named by the party whose case their testimony is likely to support. They may even be hired for just such employment. Suppose, say, a court of law is called on to decide a case that hangs on the degree of risk entailed in an activity such as running a railway. The effect of normal legal procedures is to transform it into a confrontation between the advocates of the disputants, each striving to establish the assessments offered by 'their own experts' and to demolish the testimony of the experts on 'the other side'. *Adversarial* processes undoubtedly force some of the more active sources of bias out into the open. But by dividing up the pool of expertise in partisan terms, they give no standing to experts who are not attached to either party, and who may thus perceive the situation from yet other perspectives. A direct clash of contrary opinions seldom clarifies or resolves a complex technical and social issue. Indeed, a world of totally polarised expertise would be entirely chaotic — a Wild West where risk itself would be the ruling demon.

The essence of my argument in this paper is that in circumstances such as these there is a continual social need for *independent* sources of *well-founded, unbiased* information about risks. One could say, of course, that such sources do not exist. Everybody, we are told is to some extent biased in their judgements. Nobody is capable of an entirely objective point of view. By the mere fact of their own membership of society, every expert has personal interests in societal matters. Eminent scientists who always travel by train have a different perspective on the risks of railway travel from those who normally go to work daily by car. Even those of us who are no longer gainfully employed are not entirely unconcerned about the fate of the corporations in which our pension funds are invested. And so on, *ad infinitum*.

But this objection is so trite that it is frivolous. It applies to every aspect of our lives. To act systematically on it would cut us off from entirely from our human contemporaries. The very possibility of social life is predicated on a reasonable level of trust in the honesty, sincerity and tacit benevolence of most of the people with whom we have dealings —

the other drivers on the motorway not excepted. But we also know that not everybody who claims to be acting for us is actually motivated by just such benevolence. So we have to use our ordinary social savvy to pick out the ones who seem the most reliable in that sense. In the end, the acceptability of risks becomes the acceptability of the experts. We have to choose the ones whose advice or judgement we feel it wisest to follow.

The trouble is that we are then faced by a choice amongst *strangers*. It is unlikely that we have a friendly neighbourhood specialist on railway signal systems, or on microbial pathogens in eggs, or endowment mortgage schemes, or computer viruses. So we have to rely mainly on their public credentials. From a professional CV such as an entry in *Who's Who?* we can glean information about their formal standing (e.g. Professor, FRS), academic qualifications (e.g. MD, DSc), work experience (e.g. toxicology research), official position (e.g. Public Health Officer), institutional role (e.g. Dean of Engineering), or corporate post (e.g. Research Manager). But we are not equipped to examine them on how much they really know of the subject. It is rather like choosing a plumber out of the *Yellow Pages*. In the end, we may find we have to rely on somebody of whom we have never previously heard and know practically nothing.

So when independent advice is most needed, the whole issue is likely to be already mired in dispute. The complexity of risk issues usually means that many of the accredited experts have already become associated, by employment, official position, consultancy or research contract with one or other of the organisations in conflict. However earnestly and scrupulously they try to exercise their scientific objectivity, they cannot hope to dispel suspicions of possible bias. Even in such narrowly technical procedures as peer review of scientific projects or publications, it has become *de rigueur* for expert advisers to declare any 'conflict of interest' that might influence their judgements. In a broader risk issue, where much valuable property and many human lives may be at stake, the web of material entanglement may spread much wider.

This is not a trivial matter. The problem of finding trust-worthy experts is encountered at every level of the social world. It arises in every 'risky' situation, from the personal medical consultation to the governmental commission of enquiry. In each instance, its solution is an essential require-ment for achieving acceptable closure. More generally this is often the key problem in dealing with the divisive issues that cut across our pluralistic society.

Until recently, however, this problem could usually be solved fairly easily. Indeed, the practical answer is so famil-iar that its underlying principle is seldom appreciated. In simple terms, society has come to rely on the professional personnel of academic science—the professors, lecturers, research fellows, etc. who teach and study in universities and research institutes—to perform this function. And it was precisely through their engagement in *non-instrumental* science that they acquired both the specialised knowledge and the social independence required to operate as trust-worthy expert informants on highly technical issues. This applies not only to the natural sciences and their associated technologies but also in such human sciences as law, economics and politics.

This is not to say that academic science is a community of saints and/or sages. Nor does it follow that it should be preserved unchanged for just that purpose. I am simply drawing attention to an indispensable social facility provided almost incidentally by this traditional mode of knowledge production. Unfortunately, this peculiar cul-tural form is rapidly losing its institutional independence —the very feature that guarantees its credibility as a source of disinterested expertise. But that is a point that will be taken up at length in later chapters.

FIVE

The Rise of Technoscience

The technological interface

The wild card in the language game about 'the place of science in society' is the one labelled *Technology*. Many people insist that this is quite, quite different from 'science'. But then it is almost invariably hyphenated with it as 'Science-and-Technology'. Indeed, many institutions and publications formerly devoted to the study, promotion, advancement, and teaching of science have nominally relaunched themselves by replacing the *S* in their acronym with *S&T*. But as we have seen, the boundary between these two social activities is almost impossible to define. For the present work, anyway, it is simpler to drop this *T* and follow the alternative terminology, whereby 'science' is deemed to include codified technical know-how of all kinds, whether or not it has a theoretical base.

Nevertheless, although technology is rapidly being incorporated into 'technoscience', it has a life of its own. Or, rather, it has a multitude of lives, for there are at least as many distinctive 'technologies' as there are distinctive 'sciences'. What these all have in common is deep roots in everyday life. As the anthropologists have always appreciated, the basic technical practices of a human society are inseparable components of its culture. A study of 'the place of technology in society' eventually finishes up as a study of that society as a whole. And of course science performs its main social roles through that medium.

For that reason, modern emphasis on the instrumental function of science is certainly not misplaced. But as we have seen, this emphasis underplays other social roles that are equally important. And an analysis of why it is becoming more difficult for science to perform those roles requires a clearer understanding of how it is actually shaped by its technological imperatives. That means we should be looking at it from the viewpoint of the actual sites where the demand for knowledge originates and where it is put to use.

In the first place, of course, technology is unambiguously *instrumental*. It is motivated by the perception of a material need and even the knowledge on which it is based is often 'taken for granted' in a way that non-instrumental academic science knowledge never is. Its purpose is to *'solve problems arising in the context of application'*. Considered as a mode of knowledge production it is the epitome of what some sociologists call 'Mode 2'. But this formulation is too abstract. Technologies cannot be separated from the social settings where they are practised. A stone axe, a tiled roof, a herbal remedy, or a printing press is not just a 'context for the application of knowledge',[1] or the 'solution of a problem'. Nor is it a natural object that happens to have been used as a tool. It is an *artefact* that successfully performs a meaningful function.

Technological knowledge is therefore typically *local*. It arises from, and is intended for use in, particular circumstances. *'How did we bridge that last river? Is that not the way to bridge the next one?' 'Hispano-Suiza racing cars always use carburettors of this type.' 'This drug I am prescribing ought to cure your cancer.'* And so on. Of course, as it becomes more scientific, it becomes more *generic* — that is, applicable in a wider class of life-world circumstances. It may suggest, or emerge from, or exemplify, some supposedly universal principle, some 'law of nature', some mathematical theorem. But a technology is not entitled to detach itself completely from its putative contexts of use. It must be equipped with a strong chain of

---

[1]   Gibbons, M., C. Limoges, *et al.* 1994 *The New Production of Knowledge*. London, Sage.

conceptual and practical *know-how* capable of anchoring it wherever it is specifically required to operate.

In addition to their overt uses, technological products usually perform other less tangible functions. They are often nodal entities in complex webs of beliefs, customs, roles and institutions. Automobiles, for example, are not just machines enabling personal mobility: they are also consumer goods, industrial products, engineering designs, fashion models, environmental eyesores, gas-guzzling monstrosities, weapons of war, health hazards, recreation vehicles and so on. They figure as 'cultural entities' or 'social constructs' at many different localities in modern society.

The 'contexts of application' for technology are thus extremely complex and heterogeneous. They are also typically 'path-dependent'.[2] Practical problems are shaped by the past history of their setting. Thus, the 'scientific' ideal of a perfect, absolutely optimal, general solution for a particular class of problems is quite unattainable. All that one can ever hope to achieve is a product that 'satisfices' — i.e. exceeds an acceptable threshold of performance in the actual circumstances.[3] A technological problem does not have a unique answer. There may be a number of different, equally good 'solutions', corresponding perhaps to different compromises between competing technical capabilities, or adapted to slightly different customer preferences. Anybody who has taken part in an argument between PC addicts and Apple Mac fans will know just what I mean!

Let me emphasize that the local specificity, cultural diversity, historical contingency and technical plurality of human artefacts does not make the knowledge required for their production 'unscientific'. Modern designs and manufacturing methods require vast quantities of conceptual understanding and systematized practice. The demand for technological knowledge often stimulates the generation of abstract networks of fact and theory. The mathematical dis-

[2]   Arthur, W. B. 1994 *Increasing Returns and Path Dependency in the Economy*. Ann Arbor MI, U of Michigan Press.
[3]   Simon, H. 1996 *The Sciences of the Artificial*. Cambridge MA, MIT Press.

cipline of thermodynamics, for example, came out of and also guided, the very practical craft of steam engineering. But fundamentally a technology is *a way of doing rather than just a way of knowing.* Indeed, to say that an artefact 'embodies' the knowledge that went into its making is literally true. As with a living body, only a very practised eye—that is, a mind already equipped with much understanding of the principles by which such an entity operates—can reverse the engineering and read this knowledge from it. This difficulty is compounded by the diversity of the working principles that are coordinated in a typical technological product. Very often, a characteristic component of a particular artefact—a cart wheel, say—is adapted to solve a practical problem in quite a different area of life—e.g. in a steam turbine. But to make this apparently simple idea work in practice, the dynamics of rotating masses has to be combined with knowledge drawn from quite different sources, such as thermodynamics, fluid dynamics, and the science of materials.

Incidentally, this multiplicity of sources is one of the reasons why technological innovation is not entirely analogous to biological evolution.[4] Both, certainly, are processes in which more or less 'blind' variations are selectively retained and reproduced. Indeed, the same Darwinian principle applies to all modes of knowledge creation.[5] But technological artefacts do not separate themselves into distinct 'lineages', akin to biological species, which cannot interbreed. The proud parents of the micro-chip, for example, include all the highly specialised and extremely diverse techniques required to purify the silicon, crystallise it perfectly, photograph and etch exquisitely precise patterns on it, dope these with impurities, attach electrical contacts to it, and so on, not to mention a detailed understanding of its semi-conducting properties and computational capabili-

[4] Ziman, J. M., Ed. 2000 *Technological Innovation as an Evolutionary Process.* Cambridge, Cambridge UP.
[5] Campbell, D. T. 1960 'Blind variation and selective retention in creative thought as in other knowledge processes.' *Psychological Review* 67: 380-400.

ties. It is as if an elephant could be hybridised with a sea lion and an eagle! In technological evolution such mythical beasts come alive.

The industrialisation of invention

The official chronicles of technological innovation are punctuated, of course, by accounts of the production of original *inventions* by lonely inventors. Such touching episodes are, indeed, part of the story. The vital spark in a novel artefact often emerges, as if by spontaneous generation, in that way. For this reason, the search for the sources of invention usually ends up in the thickets of personal creativity. Or it is trapped in the circular doctrine that a needed invention will be mothered by bare necessity. Or it takes for granted that technological innovations are always levered by crude economic motives, ignoring the intellectual, aesthetic and affective obsessions that so often drive their originators.[6] These, and many other psychological and social factors are surely at work, and vary in force from era to era and from culture to culture. Our modern pluralistic society certainly encourages them to the full.

But a closer reading of the historical record shows that even such an ingenious and practical invention as the zip fastener was not applied without substantial *development*.[7] That is, it had to be considerably modified, combined with other devices, reintegrated into a *prototype*, tested for defects, redesigned for ease of manufacture, and so on. Only after taking further pains — and incurring heavy costs — do practical inventions actually reach their intended users. The technological mode of knowledge production thus relies heavily on *bricolage* — the French term for the art of assembling miscellaneous entities (*'bric-à-brac'*) into a useful or beautiful object — and 'tinkering' with it until it works.

---

[6] Pacey, A. 1992 *The Maze of Ingenuity: Ideas and Idealism in the Development of Technology*. Cambridge MA, MIT Press.

[7] Ziman, J. M. 1976 *The Force of Knowledge*. Cambridge, Cambridge UP.

The development of an invention is thus a lengthy process involving the labours and savings of a great many people. Until the time of the industrial revolution this social process was not undertaken systematically. New technological concepts emerged, were tried out in practice, and evolved more or less spontaneously in innumerable anonymous workshops and craft guilds. Although a great deal of knowledge was thus being produced and put to very good use, it was fragmented and only partially codified.

But the advent of large-scale industrial manufacturing required a more systematic approach to innovation. The firms that made steam engines, for example, had to compete with one another in terms of efficiency, safety, purchase price and running costs. This induced them to incorporate all the latest inventions into their designs, to test them out and to seek novel solutions to the problems that arose. Although a firm might rely heavily on the inventive genius and inspirational leadership of a James Watt or a Robert Stephenson, its employees had to work as a team. The knowledge embodied in their products was produced collectively, in an organised manner, by a whole group of designers and craftsmen, not to mention the salesmen and bankers who kept the enterprise afloat financially.

From then on, technical 'progress' was partnered and driven by commercial forces. The trend was almost always towards more complicated and expensive artefacts or at least to ones that could be produced and sold in greater numbers. Firms that could successfully design, manufacture and market these expanded vastly. Economic growth, the mainspring of modern society, thus became completely dependent on technological innovation. Conversely, the production of technological knowledge fell largely into the hands and service of industrial capitalism.

State technoscience

The skilled personnel and elaborate facilities of modern technoscience are now mainly concentrated in the corporate sector of the economy. But a large amount of instrumental research and development is also undertaken in the state sector. In the late nineteenth century, governments began to set up institutions to regulate and foster technological progress in manufacturing, agriculture, transport, and so on. State bureaucracies needed scientific knowledge to discharge their responsibilities for the health and prosperity of their citizens, not to mention the capacity to tax them and to send them to war. Systematic research was required to produce this knowledge and to develop material and social technologies to apply it.

The work that goes on in all those 'Federal Bureaus', 'Research Establishments', 'National Laboratories', etc. might be called *state technoscience*. Like corporate technoscience, this is directed towards practical goals. But its products are not for sale. They are 'public goods' such as environmental protection, regulatory standards, disease control, weather forecasts, social security benefits, health services, weapon systems, demographic data, agricultural techniques, educational methods, etc. So the value added by these institutions is quite impossible to assess in monetary terms. Indeed, in fields such as medicine they typically operate in parallel with 'not-for-profit' organisations such as charitable hospitals and foundations.

State technoscience also differs from corporate technoscience in that it is carried out by 'civil servants'. That means that it is organised along more rigid bureaucratic lines than most commercial firms. One might think that this would allow researchers little latitude in the projects they work on. Paradoxically, however, this is not the case. State technoscience is often close to academia in its employment practices. The traditional system in most countries is for government scientists to hold tenured posts to which they are appointed or promoted on the basis of their personal scientific achievements.

What is more, since the knowledge produced by state technoscience is not usually embodied in marketable products, it is of limited commercial value as intellectual property. Indeed, the 'public goods' it relates to are often so intangible, or of such questionable practical use, that it scarcely differs from non-instrumental science in its form, substance, objectives or outcomes. In many countries, therefore, the national research organisations are closely connected with the universities and customarily produce a great deal of knowledge in the academic mode. Thus, for example, the Laboratory of Molecular Biology at Cambridge, famous for its discoveries in basic science, is an organ of the UK Medical Research Council, whose official function is ultimately utilitarian.

The modern trend, however, is to force state technoscience into the same managerial mould as corporate R&D. Their traditional mixture of civil service bureaucracy and academic individualism is considered very inefficient. It lacks — so we are confidently informed — the spur of competition for resources, the discipline of the financial bottom line, the challenge of regular performance assessment, and the acid test of market success. So it has become politically fashionable to 'privatise' as many of them as possible, despite the fact that their principal mission is to produce public goods. Even those that are still owned by the state have to be much more accountable to committees of officials and politicians — in effect, 'boards of directors' — for the implementation of their 'corporate plan'.

For the rest of this chapter, therefore, I shall assume that the state institutions that do technoscience are as much part of 'industrial science'[8] as the private sector firms with whom they are compelled to compete for project grants and research contracts. This generalisation is much too glib. It lumps together a great diversity of organisational structures and career patterns. But it will serve for the moment, at least until we get to understand the place of technoscience as a whole in a pluralistic society.

Technoscience as a mode of knowledge production

Technoscience in the fullest sense of the word only emerged in the twentieth century. By then, industrial firms had

---

[8]   Ziman, J. M. 1984 *An Introduction to Science Studies*. Cambridge, Cambridge UP.

become accustomed to recruiting scientifically educated employees from academia. As we have seen, this is still their main source of trained researchers and provides them with much of the scientific knowledge underlying their techno-logical activities. But then some of the largest firms in the electrical and chemical industries bethought themselves to close the gap between 'discoveries' and their 'applications'. Hungry for original ideas and novel findings that might be exploited technologically, they set up their own inhouse laboratories to search for them.

Ostensibly, these laboratories undertake open-ended research projects producing non-instrumental knowledge. But they would not be supported if they could not be justi-fied commercially. In practice, they are linked by internal transfers of information and expertise to the technological research, development, design and marketing of value-laden artefacts, which is the firm's real business. On occa-sions these linkages can be somewhat notional. Their wealth-creating capabilities don't stand up well to conven-tional accountancy scrutiny. So when financial sacrifices have to be made, basic research is usually the first segment of the corporate organism to be shed.

Nevertheless, the frontier between the domain of 'discov-ery' and the world of 'applications' is now wide open. This has revolutionised the process of technological innovation. Modern technoscientific artefacts are incredibly complex. They are created by enlisting and correlating a great variety of epistemic and practical resources. Technoscience is 'scientific' in that it mobilises, utilises and produces great quantities of organised knowledge — theoretical concepts, databases, research projects, experimental results, observa-tional techniques, computational models, literature searches, engineering designs, production processes, patent specifications and so on. But it is also 'technological' in that it embodies innumerable technical practices and facilities, and has clearly defined material goals, such the cure of a disease, the improved manufacture of ingenious artefacts or the means for destroying cities.

What this means is that technoscience is necessarily 'social'. It doesn't just adapt itself to, or exploit, the needs and aspirations of society. It conforms to and constructs the cultural practices and institutions where its products and know-how are employed. A personal computer, for example, is not simply a tool that we use to write texts or do book-keeping: it is a machine that has been designed — i.e. 'socially constructed' — to use us. Notoriously, it includes bundles of software that induce us almost irresistibly to purchase other products of the same manufacturer. More benignly, it shapes our language by 'correcting' our spelling and favouring certain grammatical conventions. These features have not been built in by accident. They were proposed, planned and incorporated intentionally by groups of people, working together as the employees of a commercial firm. In that process, the purely epistemic and material factors are largely subordinated to social considerations and taken-for-granted life-world concerns

To assemble all these elements, coordinate their activities, and direct them towards a specific practical goal is clearly a vast human enterprise. It can only be undertaken by a very large organisation capable of employing and managing the work of many thousands of qualified research scientists, engineers, technical personnel, craft workers, office workers, information specialists, lawyers, accountants, sales staff, etc. In a pluralistic society, technological innovation is considered to be the responsibility of private enterprise. So technoscience is mostly concentrated in the corporate sector of the economy. I am not questioning (at this stage of the argument, anyway!) the virtues of the free market economy. All I am saying here is that the usual setting in which science performs its instrumental social role is a large, often multinational, company riding on a wave of technological progress and economic growth.

The epistemic, organisational and cultural features of technoscience are shaped by this institutional environment. Take, for example, the defining characteristic of 'science' — the systematic production of codified knowledge. For technoscience this is not the object of the exercise. New knowledge is certainly being generated, but it is only valued for its contribution to the prosperity of the firm. If it cannot be exploited internally, it is designated 'intellectual property' and put on the market. But of course, once it becomes public it ceases to be a saleable commodity. So corporate technoscience renounces one of the major traditions of science. As the story of the mapping of the human genome clearly shows,[9] it has an almost irresistible urge to shroud its findings in commercial secrecy, or shackle their use by tightly-linked patents. This is perfectly legal, and is consistent with the doctrines of free market economics. But it severely limits any non-instrumental social role that this knowledge might possibly perform.

In effect, technoscience normally draws more from the general accounts of public science than it pays back into them. Of course scientific theorising is enormously indebted to the detailed investigations: serendipitous observations, powerful techniques, contrived circumstances, insightful inferences and instructive experiences gained through the systematic practice of science-based technologies. The history of science offers many examples of apparently open-ended research motivated by the possibility of achieving tangible practical goals. Astronomy, so we are told, was the daughter of navigation, and human biology was always presumed to have medical applications. Technoscience has many trading posts in these 'pre-instrumental' realms. It is particularly interested in general theories that have been 'finalised ' — that is, so well established that they can be used confidently to solve specific technical problems with determinate goals.[10] But the stakeholders of 'big technoscience' are chary of investing in very long-term ventures for exploring and colonising new fields of knowledge that might never pay a profit. Notice, for example, that the commercial rush into molecular genetics did not begin until the biological functions of

[9]   Olson, M. V. 2002 'The Human Genome Project: A Player's Perspective.' *Journal of Molecular Biology* 319(4): 931-42.
[10] Jagtenberg, T. 1983 *The Social Construction of Science.* Dordrecht, Reidel.

DNA had been well established by academic research. Sheer curiosity has no leverage on its operations.

In any case, technoscientific knowledge has to be mainly concerned with the performance of specific artefacts in particular circumstances. And yet, unlike traditional technology, it is not integrated into the life-world where these artefacts are actually put to use. Real 'contexts of application' are in the hands of patients, consumers and 'practitioners'. But even then, they are not vacant lots, patches of wilderness, empty technological niches just waiting to be occupied by brilliant innovations. To a large extent they are already staked out and virtually occupied by hungry, rival commercial firms.

Thus, for example, the life-world demand for personal transport is effectively 'owned' by the manufacturers of automobiles and aircraft. Their customers have the ultimate power of the purse. But consumer needs are typically exploited, manipulated and distorted — for example, by mass advertising — to satisfy corporate needs for profitable investments, financial growth, market shares, etc. Similarly, the demand for 'instant' food has effectively become the domain of the firms that buy it in, process it, pack and retail it. Through their market power and the other devices they appropriate the supposed 'problems' and usually end up in possession of the innovations that 'solve' them. This form of 'property' is not protected by law. Indeed, the political desire to limit its consolidation motivates the statutory control of commercial monopolies and company mergers. Nevertheless, it is an ever-present feature of our pluralist culture, and a highly significant aspect of the role of science in it.

Again, this is not the place for a debate on the push-me-pull-you dynamics of 'consumerism' and 'producerism' in the modern world. All I am suggesting is that it is naïve to suppose that the 'problem areas' which provide contexts for the application of technoscience are like unoccupied tracts on an open frontier for invention and innovation. The unexploited territories are rare, and require much patient toil to yield a living to the lone early pioneer.

Even then, the technical workers in the R&D division of the microchip manufacturer or the pharmaceutical firm are only tenuously connected with the computer salesclerk or dispensing chemist. Even when a 'problem' or a 'need' seems obvious in the everyday world, it takes a lot of defining and retranslation before it can be expressed in scientific terms. One of the achievements of scientific medicine, for example, was the recognition of distinct diseases within the jumbles of symptoms reported to doctors. Only then could they be tackled systematically.

So the 'problem areas' within technoscience itself are not marked out on the schedules of everyday life, or even on the sales charts of multinational companies. Nor do they coincide with the established disciplines and evolving specialties of academic science. Sometimes, of course, they conform to natural categories — for example, a disease such as smallpox caused by a very specific microorganism. But very often they relate to artificial entities, such as a particular type of aircraft engine or a novel system of plant breeding. The knowledge required to cover such an area has to be cobbled together out of material — and people — from a variety of sources. To make any progress, expertise has to be combined from a number of very different scholarly and practical traditions.

Multidisciplinary teamwork is thus an essential feature of technoscience. Take a hypothetical project for the development of an artificial heart. On the purely technical side, disciplines as different as electro-mechanical engineering, materials science, hydrodynamics, protein chemistry and neurophysiology are likely to be called on. But attention to the human and social aspects of the proposed innovation will require inputs from psychiatry, bioethics, social welfare, health economics, etc. This diversity is unavoidable. The final commercial success of the project will depend on open, early and skilled access to the best of what is known, or can plausibly be found out, in these very different specialised fields.

In practice, it is not always feasible to assemble such a team on a single site. For that reason, modern technoscience is heavily dependent on 'networking'. Much use is made of electronic communication systems such as the internet, both to keep team members in constant touch with one another and to gain access to external databases. But this is not peculiar to industrial R&D or to instrumental science in general. Interpersonal two-way communication is an essential component of the scientific enterprise in general.[11] All modes of knowledge production are facilitated by teamwork and 'networking'. Email is just the latest, most instant means for doing what the handwritten letter and printed journal accomplished, to very good effect, more than three centuries ago.

The fruitful direction of a heterogeneous team of highly individualistic professional scientific workers is a daunting task. That is why the literature on industrial R&D is so strongly focussed on the arts of leadership and management. What this literature does not always reveal, however, is that most such working groups last for only a few years.[12] Whether or not the proposed innovation catches on, technological progress soon requires the formation of new teams with new missions and correspondingly different members. In effect, a successful career path in technoscience is seldom confined to a permanent problem area. It tends, rather, to wander from project to project, wherever the relevant expertise is needed. At the same time, the continuous creation of new teams requires a well populated and varied pool from which to draw.

This type of 'matrix management' is only possible inside a very large organisation. This is another of the reasons why technoscience is mostly concentrated in major commercial corporations, employing people with a wide range of expertise. This is not really inconsistent with the well-known fact that a considerable amount of the innovation in some industries — notably pharmaceuticals — originates in quite small firms. In reality, however, so-called 'research bou-

[11] Ziman, J. M. 1968 *Public Knowledge: The Social Dimension of Science.* Cambridge, Cambridge UP.
[12] Ziman, J. M. 1987 *Knowing Everything About Nothing: Specialization and Change in Scientific Careers.* Cambridge, Cambridge UP.

tiques' specialising in advanced technoscience are very unstable. A few survive by either growing large enough to diversify their activities or by subordinating themselves contractually to one of the existing giants.

Indeed, the practice of 'outsourcing' research to quasi-independent satellite firms is characteristic of post-modern management. Monstrous monolithic all-singing, all-dancing bureaucratic corporations are quite out of favour. Big industrial firms nowadays buy in many of their services — including much that would previously have been the responsibility of their research and development divisions — from outside contractors, and devolve responsibility for the development of new activities to temporary *ad hoc* 'teams' or 'working parties'. This enables them to respond rapidly to changing market conditions and/or novel technological opportunities. Many of the typical organisational features of modern technoscience — for example, flexibly configured, global webs of diversely funded, transient, heterogeneous workteams (to encrypt a very complex picture!) — merely reflect current managerial practice in the corporate sphere. In this respect, 'Mode 2' knowledge production[13] is just the 'post-industrial' version of our old friend industrial R&D.

From a corporate point of view, the beauty of this quasi-market system is that it enables them to shed staff quickly when things go wrong. Fortunately, when the skids are pulled out from under a small technoscience enterprise, its employees don't necessarily suffer greatly, for they usually have valuable personal expertise to sell. But they do have to become accustomed to moving from firm to firm as new capital is ventured and new project teams are assembled.

From an evolutionary perspective, of course, the high mortality rate for small technoscience units is a sign of progress. Their fate is tied to the commercial success of their products. Whether or not these are sound in principle, if they do not sell, then no lunch! This is as it should be. Tech-

[13]   Gibbons, M., C. Limoges, *et al.* 1994 *The New Production of Knowledge.*
       London, Sage.

nology evolves by the 'selection' of just a few novel artefacts out of numerous other 'mutants'.[14] In effect, technoscience is a risky business, rather like oil prospecting. Drilling is ferociously expensive and most wells are dry. Once in a while an innovative product, such as viagra or the microchip, produces a bonanza. But it is essentially a lottery, where a vast amount of capital is needed to be sure to have bought a regular supply of winning tickets.

The trouble is that the knowledge that has gone into the making of all these hopeful innovations is not systematically collected or evaluated. The test of market success is purely pragmatic. Yes, the new gizmo operates perfectly in the context of application, but does that mean that its supposed working principle is valid more generally? As I have said, science in the large benefits immensely from the mere fact that such devices do sometimes, eventually, perform as planned. But technoscience itself takes no interest in how a successful practice ought to be interpreted more widely.

And what about all the know-how embodied in the gadgets that didn't take off commercially? From an epistemological point of view, these are just like 'failed' experiments. They effectively 'disconfirm' the theories or supposed facts on which they were designed. Much truth might be learnt from public accounts of how they went wrong. But of course, as the frenetic secrecy surrounding the actual results of clinical trials of pharmaceutical products indicates, this is information of high commercial value. The self-critical attitudes that are vital inside each corporate wall must never leak out. It is not even customary to 'knock' directly the work of rival firms, for this might throw doubt on the products of the whole industry. Technoscience is not a product of conceptual falsification *à la* Popper. It actively discourages sceptical attitudes, especially towards its own institutions and practices.

---

[14] Basalla, G. 1988 *The Evolution of Technology*. Cambridge, Cambridge UP, Mokyr, J. 1990 *The Lever of Riches*. New York NY, Oxford UP, Ziman, J. M., Ed. 2000 *Technological Innovation as an Evolutionary Process*. Cambridge, Cambridge UP.

In practice, these considerations have little direct impact on 'laboratory life'. In that restricted world, most 'problems' are presented in narrowly technical terms.[15] This enables individual scientists to exercise 'technical autonomy' in the way that they set about solving them. This is how their professional expertise is mostly demonstrated and for which it is rewarded. For those who are good at it, there is nothing like problem-solving as a vocation. That is the kick that the great majority of scientists get out of their work, and technoscience provides abundant opportunities for putting on one's boots

Indeed, in spite of all their talk about freedom in research, it seems that most scientific workers attach limited personal importance to 'strategic autonomy' — that is, the right to formulate the projects and programmes in which their capabilities will actually be enlisted.[16] This is just as well, for technoscience entrusts this role to many other people, most of whom are not very knowledgeable scientifically. The final responsibility for what is done in a commercial firm rests with its 'proprietors' — the senior managers, executive directors, shareholders, etc. They pull the levers of the bureaucratic machine that sets the goals, approve the projects, monitors the performance and disposes of the results.

Creative discoverers, inventors and designers are the sparks that actuate the great social engine of technological innovation. They are certainly respected, listened to and honoured in retrospect. But they have to work harmoniously with people with other expert responsibilities such as profitable marketing. At times they have to undertake, and perhaps lead, projects and programmes whose goals seem quite unattainable. At other times they must hurry to complete work whose outcomes are still very insecure. And they must operate within complex institutional frames where they are often quite lowly members of steep bureaucratic hierarchies. Occasionally they are advanced to

[15]   Latour, B. and S. Woolgar 1979 *Laboratory Life: The Social Construction of Scientific Facts*. London, Sage.
[16]   Jagtenberg, T. 1983 *The Social Construction of Science*. Dordrecht, Reidel.

positions of high managerial authority. But specialised 'problem-solving' expertise is not automatically given precedence over other personal skills, such as technical 'trouble-shooting', project management, financial dexterity, salesmanship, etc.

What this means, above all, is that the incoming scientists have to conform to the epistemic culture of commercial technoscience. They are employed by a specific company to work for it, not for the advancement of learning nor the welfare of society. Although this work typically involves the generation and transmission of knowledge claims, the absolute validity of these claims is not an over-riding concern. They only have to satisfy internal criteria of 'truth'. All that is required is that they be considered sufficiently reliable for the job in hand, good enough to sustain a proposed commercial venture, and as convincing as seems necessary to persuade the directors not to go ahead after all.

Well, that is not really much below the status of much that normally passes for well-founded scientific knowledge.[17] The trouble is that internal scientific findings sometimes clash directly with the established policy of the company, and that what then seems to be truth does not necessarily prevail. Their contract of employment does not require the employees of a firm to lie actively on its behalf. But they understand perfectly well that they put their jobs on the line if they do not toe the line, or at least keep silent, on matters where corporate interests are at stake. Their behaviour is then sadly torn between normal life-world prudence, and transcendental epistemic morality.

The social role of commercial technoscience is to provide society with a great variety of novel products. Besides tangible and intangible artefacts, these include design concepts, manufacturing techniques, bodies of knowledge, and skilled expertise. What these all have in common is that they belong legally to the firms that produce them. The right to own these products, to control their use by others, and to

---

[17]   Ziman, J. M. 2000 *Real Science: What it is and what it means*. Cambridge, Cambridge UP.

dispose of them for cash, is basic to our market economy. Faced with a practical problem, we are expected to seek out a firm that offers a practical solution, and to buy it from them. In principle, our need will already have been anticipated, and the technoscience required to meet it will already have been understood, ideally by several competing firms.

How technoscience benefits from non-instrumental science

In the previous chapter we looked at the various ways that society benefits from science. The bulk of these benefits are directly material. They were the expected outcomes of well-conceived projects. The dominant mode of knowledge production in modern society is highly instrumental — that is due to the technoscience powering capitalist industry. It can be a cruel and messy way of putting knowledge to work, but it produces the goods. We don't get from it all that we need, and what we do get is not always what we need, but it is fabulously productive.

On the other hand, many of the benefits of systematic knowledge production are so broadly distributed that they could not have been preconceived and generated intentionally. Some of them, it is true, are side products of instrumental research. For example, a large part of our understanding of the nature of the human animal is essentially a 'spin-off' from medical technoscience. Many of the wonders of modern astrophysics and planetary science were revealed by research designed specifically to aid navigation, or military weaponry, and so on, in innumerable particular ways.

At the same time, a great deal of scientific knowledge is still produced 'non-instrumentally', with no intended practical application. That does not mean, of course, that this apparently 'useless' knowledge is entirely valueless, even from a strictly utilitarian point of view.[18] As we are told again and again, some of it will surely have extremely valu-

---

[18] CSS 1989 *The Value of Useless Research*. London, Council for Science and Society.

able direct applications that we cannot specifically foresee. Therefore, so the argument goes, its production now *really* has a 'pre-instrumental' function. For example, very large governmental expenditures on basic biological science are easily justifiable as strategic moves in the never-ending war against disease. The prudence of this policy need not be questioned.

As I tried to show in the previous chapter, even research with no conceivable practical applications — cosmology, say, or prehistoric archaeology — is of societal value. Its intangible gifts to humanity and to the social order vanish from sight in a world awash with material interests, sunk costs, financial bottom lines, intellectual property rights, practical outcomes, 'can do' politics, economic competitiveness, performance indicators, commodity markets and all the other utilitarian signifiers of our modern culture. And yet it can play an important role even amongst the dark satanic mills of corporate industry. The mighty engine of technoscience that powers the industrial economy cannot prosper without the other modes of knowledge production practised elsewhere in society.

Quite apart from its obvious 'pre-instrumental' role, the non-instrumental science traditionally undertaken in academia, makes vital indirect contributions to instrumental technoscience. Experienced managers of industrial R&D often argue, for example, that they cannot operate satisfactorily in a country without strong universities and other academic institutions. Although this fact is often overlooked by politicians, publicists and economists, it is one of the characteristic features of our pluralist society.

The most obvious benefit, of course, is a supply of trained researchers. This is not just a matter of instructing would-be scientists in the theories, facts and techniques that they will need when they enter industrial employment. It requires them to gain personal experience of the whole research process, from the germ of an original idea to the ripened fruit of a well-established item of knowledge. Technoscience depends for its continued vitality on a regular inflow of 'self-winding' scientists accustomed to considerable autonomy in performing open-ended projects. In principle (if rather imperfectly in practice) this is precisely the ultimate goal of doctoral and postdoctoral education, away from the pressures of practical application.

Needless to say, instrumental technoscience calls upon a wider and deeper body of knowledge than it is able or willing to directly produce itself. This need is much more general than simply to provide access to potentially exploitable discoveries. It includes, for example, the broad-brush world pictures in which so much modern science and technology is initially sketched out. It includes speculative knowledge that has been made reliable enough to form a basis for practical application. And even commercially-oriented technoscience often lacks direct contact with the life-world conditions of its ultimate customers. Thus whole chunks of knowledge from the human sciences are needed to provide realistic perspectives on future social needs.

Moreover, as we have seen, technoscience is always a risky enterprise. It is not just commercially or politically uncertain. It often produces quite unintended adverse effects on the public. But the identification, assessment and disarming of such risks requires technical and societal knowledge from outside the instrumental frame of the proposed development. For example, the whole modern discourse on the ethics of science, engineering, medicine, etc. is based essentially on academic scholarship in the human sciences. In effect, technoscience would be socially unacceptable without a thriving system of non-instrumental science to curb its excesses.

Finally, even the most utilitarian research enterprises often get into situations where they desperately need knowledgeable but disinterested advice. For example, there may be institutional deadlock over a controversial technical issue, or serious concern about the viability of a major project. For elementary commercial or political reasons, this little local difficulty may have to be kept secret. A natural way of dealing with it is then to obtain a confidential opinion from an impartial 'outsider'. Thus, technoscience benefits as much from its independence as from the expertise of the specialist consultants and advisers that it frequently calls in from the world of non-instrumental research.

Conditions for non-instrumental science

In the next chapter we shall look at 'academic' science, the traditional non-instrumental mode of knowledge production. But first let us consider a rather obvious question. Couldn't technoscience itself provide all the benefits that society has previously received from the 'not-for-profit' sector of the scientific enterprise? Shouldn't we just accept that 'Mode 2' now totally supersedes 'Mode 1'?

The trouble is that the non-instrumental roles outlined in the previous chapter can only be performed under certain general conditions. Unfortunately, these conditions are almost the contrary of those that prevail in modern technoscience. To see this, let us look briefly at what is actually required of non-instrumental science to obtain its social benefits.

In the first place, the knowledge it produces is needed for open use in law, politics, and social issues. In other words, it must be completely *public*, from the moment that it takes definite shape. Thus, for example, the credibility of an expert witness evaporates as soon as he or she lays claim to secret knowledge that cannot be challenged by adversarial cross-examination. The requirement for transparency is incompatible with the basic principle of commercial technoscience, where knowledge is deemed to be private property, and may only be disclosed with the permission of its legal 'owner'.

Again, many of the non-instrumental functions of scientific knowledge require it to be *universal*. That is to say, it must be broadly applicable and able to be made generally accessible. This is essential if it is to contribute to a scientific 'picture' of some aspect of the world, such as the origins of life on earth or the sources of technological innovation. Technoscience, by contrast, is primarily concerned with particularities. Its principal function is to provide specialised elites and specific groups with information needed in localised contexts.

To remain fully open to knowledge about the natural and social worlds, science must also be *imaginative*. It must be curious about every aspect of life and being, and ready to speculate about what might be found in previously uncharted realms. This is very far from the spirit of technoscience, which is tied to what can reasonably be achieved through what is already adequately understood. It is fully occupied with solving perceived 'puzzles' within established paradigms,[19] and can spare no thought about overarching conceptual problems or scarcely recognized difficult enigmas.[20]

A particularly valuable non-instrumental feature of science is its capacity for *self-criticism*. This not only ensures that scientific conjectures are put through the mills of experiment and debate. This attitude plays a central role in our pluralistic culture. It fosters systematic questioning and rational justification of all stated purposes and potential achievements, as they relate to the goals and actions of others. Technoscience, on the other hand, is sensitive only to the pragmatic tests of the marketplace. It cannot stomach any analysis of the social consequences of its own products, or of the activities of its corporate proprietors.

Finally many of the non-instrumental roles of science depend on its reputation for *objectivity*. Philosophically speaking, this is an unattainable ideal, but it betokens sufficient 'independence of mind' to warrant trust in many contentious social situations. Technoscience can make no credible claim to this virtue. In its material operations, certainly, it must satisfy the life-world realities of physical objects and biological organisms. Its aircraft must fly, and its heart surgery must cure. But in its social relations it is harnessed to the interests of its 'owners', and is no more

---

[19]   Kuhn, T. S. 1962 *The Structure of Scientific Revolutions*. Chicago IL, U of Chicago Press.

[20]   Ziman, J. M. 1981 *Puzzles, Problems and Enigmas: Occasional Pieces on the Human Aspects of Science*. Cambridge, Cambridge UP.

trustworthy than these allow. By its very nature, it is bound to be partisan.

This brief summary of a number of extremely complex issues is obviously very schematic. Nevertheless, it is clear that the conditions under which science customarily carries out its non-instrumental roles are not provided by technoscience. In effect, they conflict directly with the way in which science and technology perform the instrumental functions that are also required of them by society. Blind confidence in the power of 'the market' might suggest that all responsibility for the production of scientific knowledge should be handed over to the corporate sector.[21] As thoughtful economists have been at pains to emphasize, this view ignores the inestimable value of knowledge as a pre-instrumental public good.[22] Sociologists and political scientists would do well to remind them also of the variety of other social functions regularly performed by scientific knowledge and scientific communities.

[21]   Kealey, T. 1996 *The Economic Laws of Scientific Research*. London, Macmillan.
[22]   Dasgupta, P. and P. A. David 1994 'Towards a new economics of science.' *Research Policy* 23: 487-521.

SIX

Academic Science as a Non-Instrumental Mode of Knowledge Production

'Academic' Science

Technoscience, whether state or corporate, cannot perform all the non-instrumental roles that society requires of science. That's not a problem. All those desirable social functions and benefits are already being provided in abundance by 'academic' science. (Hereafter I shall drop the quotes around this word. But I have put them in here to indicate that I am not just talking about the research that happens to be done in universities or under the auspices of national academies. It also includes a great variety of other social institutions where scientific activity is undertaken on similar terms.)

The essence of academic research is that those who perform it are deemed to do so in a personal capacity. To use the revealing phrase that is customary in university circles, it is 'their own work'. They hold posts that they have won through their personal contributions to knowledge, as assessed by their scientific 'peers'. They are confidently expected to gain further scholarly esteem by continuing along that path, and are provided with the resources to do so. But they are not formally required to carry out any specific research projects—indeed, in some institutions they may go on picking up their pay cheques until they retire on a comfortable pension, even if they do no further research.

We can see, in principle, that established academic scientists are supposed to behave as if they were just gifted amateurs and so dedicated to the pursuit of the knowledge goddess that they run madly after her just for the honour of having touched the hem of her gown. The sordid concept of research as a paid profession is carefully avoided. The whole idea is to find people who have shown some aptitude for this noble calling and then give them the material means and social standing for them to make it their life work.

The traditional arrangement is for academic scientists to be employed officially as lecturers, even though they spend most of their time doing research. This is a convention that arose in the German universities early in the nineteenth century and is now followed all over the world. But there have always been a few non-teaching posts for notable scholars in state-supported 'National Academies', and many countries nowadays devote substantial public funds to scientific work under similar conditions in their 'research councils', 'national research centres', 'research organisations', etc. Some universities, also, now offer more or less permanent research posts that are free from teaching responsibilities and yet are not tied to specific research programmes.

In other words, this is a scientific culture whose ethos is not seen as being explicitly instrumental. In principle, an academic scientist is free to tackle problems whose relevance to practical life is entirely beyond rational conjecture. Of course, she may actually choose to work in a field that is close to a known 'context of application', or to develop a 'strategic' understanding of a body of knowledge with a realistic potential for ultimate utility. But such quasi-instrumental or pre-instrumental considerations are simply not to the point. All that matters to the institution or academy is whether the research yields, or may reasonably be expected to eventually yield, a 'contribution to knowledge'.

Needless to say, that last sentence is alarmingly weak. Almost any new fact or hypothesis, however feeble, would

seem to fit that description. We are reminded of the derog-
atory sense of the word 'academic', meaning 'abstract',
'unrealistic', 'hypothetical', and so on—in short, very near
to absolutely useless! By comparison with sturdy instrumental terms such as 'profitable' or 'feasible', it seems very
vague and ineffective. In practice, as we shall see, although
academic science is studiously non-utilitarian, it applies
other powerful criteria to what it accepts as 'reliable knowledge'. Nevertheless, there is a genuine problem, to which
we shall return, of just how much we can actually rely on
research claims that have not been put to the test of
technological application.

For the moment, the main point is that the ethos of academic science fits it perfectly for its non-instrumental societal roles. This, of course, is no accident. Academic science
co-evolved with the other features of our pluralistic socio-
economic culture. Its methodological principles and communal practices were not handed down from on High by a
Supremely Rational Authority; they emerged and became
established along with the political, legal and commercial
institutions with which it is connected. And it is adapted
closely to the changing needs of increasingly powerful technologies, even though it is always seeking to escape their
grip. On that matter, we shall report more later on.

Academic science as a social institution

The academic way of life is a distinctive cultural form that
has spread nearly unchanged throughout the world. Yet
those of us who are immersed in it, like fish in water, often
overlook its peculiar features. These were famously
summarised by Robert Merton, more than half a century
ago, as a set of social norms.[1] But quite a lot of sociological
and philosophical explication is needed to show how
these abstract principles — *Communalism, Universalism,
Disinterestedness, Originality* and *Scepticism* — operate in

---

[1]   Merton, R. K. (1942 [1973]). The Normative Structure of Science. *The
Sociology of Science: Theoretical and Empirical Investigations*. N. W.
Storer. Chicago IL, U of Chicago Press: 267-78.

practice. Deeper analysis is then required to understand how the community regulated by these '*CUDOS*' norms actually holds together and manages to produce a particular kind of knowledge.[2] And in the end, as with all cultural forms, we are still faced with great enigma of the human sciences: how is it that supposedly self-centred individuals can be induced to combine constructively in collective social enterprises?[3] So let me leave out the theorizing for the time being, and keep to observable facts.

To appreciate just how well academic science is suited to its non-instrumental social functions, let us look at some of its routine practices. Take, for example, the worldly precept 'Publish or Perish!'. This reminds researchers that they only get professional credit for the work that they publish, and thus ensures that the knowledge produced by academic science is opened up to the world. As we have seen, this *transparency* is a vital factor in all the non-instrumental roles of science.

Again, however arrogant academics may become as they advance through their hierarchies, they owe their initial employment and early preferment to their proven capabilities as scholars and researchers. That is, academic science is a strongly *meritocratic*. It is 'a career open to the talents'. In principle, its members are not associated with any particular class, ethnic group, political party, religion or other sectarian social movement, and are thus free to serve the interests of the whole population. In practice, of course, differential access to higher education means that this universalist ideal is seldom attained. That is why the relative lack of women, or the unbalanced representation of minority ethnic groups in science, is such a serious issue. Nevertheless, this feature of academic science is essential to many of its non-instrumental functions.

---

[2]   Ziman, J. M. (2000). *Real Science: What it is and what it means.* Cambridge, Cambridge UP.

[3]   Ziman, J. (2002). No Man is an Island. *Hermeneutic Philosophy of Science, Van Gogh's Eyes, and God: Essays in Honour of Patrick A. Heelan, S.J.* B. E. Babich. Dordrecht, Kluwer: 203-18.

Observe also that it has always been a 'global' calling. Scholars are notorious as migrants, with brains that are too easily drained from country to country. Even when firmly rooted in their home lands, they delight in meeting their intellectual colleagues in transnational 'invisible colleges' where they can talk over the same highly specialised problems in their shared technical tongues. Indeed, the charge of 'cosmopolitanism' levelled at them by populist nationalists shows that they are not the tools of any one state, and symbolises their value to humanity at large.

Another academic tradition, now somewhat eroded, is 'tenure'. In its most extreme form, this meant election at a relatively early age to a permanent job for life, or at least up to the age of a well-pensioned retirement. What is at stake here is intellectual *autonomy*. Protected thus from extraneous material concerns, individual scientists are free to undertake research on problems of their own choosing, and to present the results without fear or favour. What is more, they are not to be compelled by threats of dismissal to serve the interests of the body that employs them as those in technoscience are. In other words, this practice enables academic scientists to claim the independence of mind required to perform many of their non-instrumental social roles.

In reality, however, individual researchers and small research groups are strongly dependent on their parent institutions for the resources they actually need for their research projects. Academic research institutions, in their turn, have to find the funds to provide these resources. In effect, they need relatively disinterested *patronage*.[4] For that reason, elaborate systems have emerged to distribute state or charitable funds to academic research projects solely on the basis of their scientific merit, and almost regardless of their practical potentialities.[5]

---

[4]   Turner, S. P. (1990). Forms of patronage. *Theories of Science in Society*. S. E. Cozzens and T. F. Gieryn. Bloomington IN, Indiana UP: 185-211.

[5]   Weinberg, A. M. (1962). 'Criteria for scientific choice.' *Minerva* 1(2): 158-71.

Above all, lay accounts of academic science stress its devotion to 'discovery'. The production of *new* knowledge is honoured far more highly than the *re*production of old hat. *Originality* is the key note of a whole spectrum of symbolic rewards, from the award of the degree of PhD to the global accolade of a Nobel Prize. It is also the principal component of the 'merit' that academic scientists require for ordinary professional advancement. More often than not, of course, the requisite novelty is somewhat notional, and perceptible only to technical experts against the background of current knowledge in a very narrow field. But academic science prides itself on being an imaginative enterprise that benefits society by positively welcoming almost any new idea, however wild or contrary to the established order, that can be made sufficiently convincing.

And there's the rub. What makes a new idea 'convincing'? Non-instrumental science is of no value to society unless the knowledge it produces is reasonably reliable. So, many of the practices of academic science are essentially *self-critical*. The best known of these is 'peer review' — that is, the scrutiny of research claims by the scientific 'peers' of the claimant before they are accepted for publication. There are also other communal practices, involving more open discussion and debate, which serve a similar purpose. That does not mean, of course, that well-established scientific knowledge is so sound that it can be treated as absolutely 'true'.[6] These procedures are not even very systematic, in the sense that not all the research findings in the scientific archives have actually been expertly assessed and found 'sufficiently convincing' to be worthy of serious consideration. But they do ensure that the knowledge produced non-instrumentally by academic research can usually be relied upon pretty well — almost as well, in fact, as much as the technoscientific knowledge — that has seemingly passed the pragmatic tests of daily use.

---

[6]   Ziman, J. M. (1978). *Reliable Knowledge: An Exploration of the Grounds for Belief in Science*. Cambridge, Cambridge UP.

It is easy to be scornful of the Mertonian norms. Certainly, they are ideals that are impossible to follow perfectly in real life. Sceptical sociologists have compiled impressive lists of cases where they have been scandalously infringed. So they can be cynically dismissed as the ideological wish-list of a complacent elite. Nevertheless, they do sum up many of the ways in which academic science differs from other social institutions and cultural forms. The historical evidence is that for nearly two centuries whole communities of research scientists have been constrained to follow, more or less religiously, something like the institutional practices that I have outlined above. During this time, of course, these practices have slowly evolved in parallel with the societal functions that they enable. In effect, they are the small print in the implicit 'contract' between our modern pluralistic society and an institution that provides it with a plurality of benefits.

Academics as experts

It is a matter of simple fact that we have come to rely on our academic institutions and their individual members for trustworthy expert knowledge, and fully informed by disinterested opinion, on many matters of public concern. For example, academic science is deeply involved in the handling of risk, which now figures so prominently on the political agenda. This involvement is so customary and familiar that we take it for granted. And yet it is a vital element in our pluralistic, democratic way of life.

How is it that 'academics' are enabled to perform these functions? It is not because they are peculiarly well-informed on the technical aspects of potential hazards or other societal issues. The engineer in charge of the leaky chemical plant, the public health official fully cognisant of the statistics, the repairs manager on the railway, are probably much more aware of what has and could again go badly wrong. The nameless researchers in certain governmental or corporate think-tanks surely have a better inside knowledge of the likely shape of many things to come. And there is always the problem of getting academics to understand some of the realities of life outside their ivory towers.

What is more, there is an immense amount of knowledge being produced by extremely accomplished experts outside of 'academic' science. Indeed, in terms of money, people and technical resources, the research and development work undertaken by industrial firms and governmental bodies dwarfs what is done under academic auspices. Much of what is studied and discovered in their laboratories is almost indistinguishable from what might be contributed to knowledge from a university department or institute. Technoscientists working on problems that have started in 'contexts of application' often find themselves exploiting — even challenging — the most fundamental theories. Conversely, academic scientists have to include the contrived properties of technological artefacts in their basic 'world pictures'.

In fact, as I have already remarked, at the level of the laboratory bench, academic science and technoscience are almost indistinguishable. The two modes of knowledge production use essentially the same theoretical paradigms, technical apparatus and research methodologies. Skilled researchers can move from one milieu to the other almost at the click of a mouse. They jabber the same jargon, read the same papers and search the same databases. Very often they work on practically the same problems, and arrive (necessarily, for they are both doing *science* aren't they!) at effectively the same solutions. So if the social role of the scientific expert were simply to be extremely 'knowledgeable' on a particular subject there would be little reason to choose between them.

The real point, however, is that they have become knowledgeable along different paths, in different contexts. Technoscientists acquire their expertise in the course of research on problems prescribed by their employers: academic scientists become learned and formulate research projects on subjects of their own choosing. In principle, they are not answerable to any other authority than the considered opinions of their scientific peers. A significant degree of 'strategic autonomy' is of the very essence of the academic ethos.

In practice, of course, graduate students, technical staff, and research associates on short-term contracts — indeed a large proportion of the individuals actually engaged in academic science — are not free to choose their own research problems. Even for a fully-fledged 'principal investigator', the options open at any given moment are limited. There is no getting around such factors as previous training, career commitments, and the availability of apparatus, not to mention the sheer feasibility of the proposed investigation.[7] Nevertheless, they do have a lot of room in which to manoeuvre, and are expected to decide for themselves how best to use their personal capabilities, intellectual opportunities and material resources.

This autonomy is not only a prerequisite for the norm of 'originality'. It is the medium through which academic scientists display the 'creativity' that is their most valued talent. It applies to their teaching as much as to their research. A serious scholar does not just write sound text-books and give well-informed lectures. She presents her *own* account of the subject, offering her *own* interpretations of contentious issues and suggesting her *own* solutions to its outstanding problems. As well as taking personal responsibility for her own research results, she is expected to have an independent opinion on the credibility of the knowledge being produced in her field. She need be constrained only by her own judgement of the effects of her words and actions on her reputation and career. In sum, once they are 'established', academic scientists operate as free individuals within their particular domains of knowledge,

This privilege is not due to inherited status, precocious talent, a sophisticated education, or hierarchical seniority. It has to be earned, over a period of years, by competent research performance. And by its very nature, it cannot be regulated by some higher authority. Yet the communal

---

[7]   Ziman, J. M. (1981). 'What are the options? Social determinants of personal research plans.' *Minerva* 19: 1-42.

norms and institutional practices that have shaped it are seldom made explicit. They are simply internalised as personal values, goals and responsibilities. So phrases such as, 'doing my own work', 'undertaking pure research', 'satisfying curiosity', or 'pursuing knowledge for its own sake' are not just self-serving cant. They reflect the practical meaning of 'academic freedom' for individuals who have become accustomed to it.

Thus, the knowledgeability of the academic expert is not just 'technical', like being able to solve the practical problems of an established craft. Nor is it just 'scholastic', like being able to recall the correct answers to a set of examination questions. It could perhaps be described as 'intimate', as of a territory that one has personally selected, explored, cultivated and made ones own. Remember that the ideal of academic life is of the ardent amateur. Having worked hard to develop the capabilities of ones private estate, one is both proud to display them and anxious for their preservation.

This feeling of personal ownership is well demonstrated by the 'reality' that academic scientists typically attribute to their theoretical concepts.[8] How could anyone doubt the existence of the invisible entities that they have been manipulating mentally to such good effect? For them, an outlandish hypothetical construct, such as a gene or a quark, can soon become as familiar and robust as the spade in the toolshed or the kettle on the hob. No wonder there is so little public understanding of their ways of thought!

And yet there is also an air of 'unreality' about academic knowledge. It is so frequently clouded in controversy. All too often the 'indisputable' facts and theories of one expert are equally convincingly controverted by those of another. For reasons to be discussed shortly, epistemic pluralism and ambiguity are intrinsic to academic science as a mode of knowledge production. The same applies amongst technoscientists, of course, but they are not so much into theorising, and can more easily resolve their disputes by the

[8] Ziman, J. M. (2000). *Real Science: What it is and what it means.* Cambridge, Cambridge UP.

use of the strong acid of practice. Streetwise academics are well aware of this innate uncertainty, and are careful to shroud their pronouncements in a mist of *caveats*. But what the general public wants is a clear cool spring of disinterested, relatively reliable knowledge. The opaque, lumpy mixture of dogmatism and agnosticism typically offered to society by academic scientists can be very off-putting. Why don't they learn to perform their part in the social drama more helpfully and gracefully?

Academic attitudes

What seems like ones personal identity as a 'scientist' is indeed a social role. But it is not scripted for everyday life: it is shaped and defined by the communal stage on which it is mainly performed. And it is precisely in this role, rather than for their individual qualities of intellect and character, that academics are accredited to society at large. Their place in our culture is to tell it as *they* see it, not as they suppose that *we* perceive the world.

As we have seen, one of the major non-instrumental contributions of science to society is a particular 'attitude'. Scientists are valued for their cool rationality, their patient respect for material evidence and their orderly style of argumentation. But what is more important is that their minds are always open to new ideas. Without being compulsive revolutionaries, they celebrate novelty, and tolerate dissent. They strive for unifying theories, yet most of their thinking exemplifies the pluralism that is actually the organising principle of modern society.

It is possible, but unproven, that it is mainly people with the appropriate psychological disposition who become scientists. It is scarcely surprising, for example, that scientists typically exhibit formal rationality, since that is akin to the mathematical ability that has opened a whole succession of educational doors for them, right up into the research world. Again, eminent scientists often report anecdotally that they were animated from childhood by insatiable curiosity, and were thus drawn into science willy-nilly. But we seldom hear about this from the many others (that may be equally eminent) who would make no such claim.

It is more likely that the requisite mental traits are fostered and developed by active experience of research. Thus, scientists learn the hard way to be receptive to novel concepts, but always to pay scrupulous attention to the brute facts that might so easily defeat their speculative forays. Thus they become pluralists perforce, not because they have given up on the high ideal of a macro-theory of everything but because they have to keep in mind the various conflicting micro-theories that seemingly explain various features of their confusing results.

Nevertheless, it is not obvious that a fully 'scientific' attitude is necessarily required for or produced by professional involvement in the production of scientific knowledge. As we have seen, technoscience tends to cultivate a technocratic mentality that cannot abide pluralism. Again, the great weakness of all instrumental science is that it favours technical virtuosity above curiosity. 'Mode 2' knowledge production[9] — networked, multidisciplinary team research on problems arising in contexts of application — although it benefits enormously from intellectual creativity, has no established place for an individual with 'an enquiring mind' and does not officially encourage that disconcerting trait.

It seems, rather, that this mental orientation is specifically associated with *academic* science. That is not, I guess, because academic science is typically non-instrumental, but because that is the attitude favoured and/or formed by its social practices. Take, for example, intellectual curiosity. This is heavily promoted by the academic norm of originality. Thus, one of the best ways to win a Nobel prize is to seize on a chance observation — the correlation between mental illness and ritual cannibalism in certain New Guinea tribal groups, the regularity of the radio signals from certain stars, the theoretical possibility of rolling a sheet of graphite into a

[9]   Gibbons, M., C. Limoges, *et al.* (1994). *The New Production of Knowledge*. London, Sage.

a tiny sphere—and probe it relentlessly until it gives up its secrets. Academic scientists are not just in the business of making discoveries: they are fascinated by anecdotes involving 'serendipity'—i.e. occasions when independence of mind and of circumstances have combined to permit the active exploitation of just such opportunities.[10]

But the most beneficial contribution of academic science to the social order must surely be its attitude towards controversy. As we have seen, scientists have to be able to take and give serious criticism of their ideas and findings without getting 'personal' about it, and breaking up into warring factions. This attitude develops through participation in a variety of communal practices. Academic communities are not just disputatious:[11] their scepticism is *organised*.[12] Although the emotional tone of these practices is carefully controlled by conventional courtesies and constraints,[13] they are not just dramatic spectacles. The actors have to play them for real, which puts them under considerable psychic strain. In the end, one might say that they internalise the mental and emotional procedures by which destructive individual conflict is systematically transformed into constructive social achievement, and rehearse in their own minds the likely flow of debate. That is a rather high-flown notion, but I think that it indicates the subtlety of the scientific attitude.

Nowhere is this attitude more important than in the actual construction of knowledge. The fundamental difficulty with all non-instrumental science is that it cannot fall back on the test of practical utility. Technoscience can say: 'This is what worked well in these circumstances in the past,

[10]   Roberts, R. R. (1989). *Serendipity: Accidental Discoveries in Science.* New York NY, Wiley.

[11]   Campbell, D. T. (1979). 'A tribal model of the social system vehicle carrying scientific knowledge.' *Knowledge: Creation, Diffusion, Utilization* 1: 181-201.

[12]   Merton, R. K. (1942 [1973]). The Normative Structure of Science. *The Sociology of Science: Theoretical and Empirical Investigations.* N. W. Storer. Chicago IL, U of Chicago Press: 267-78.

[13]   Ziman, J. (2000). 'Are debatable scientific questions debatable?' *Social Epistemology* 14(2/3): 187-199.

so you can rely on it to work just as well now.' The criteria by which academic science claims to be reliable are much more complex and questionable.[14] But they are not abstractly 'philosophical': they are implicit in the critical social practices that permeate the whole research process.

Academic science is centred on its public 'archive'.[15] This is scarcely surprising, since it is mostly undertaken in universities. In academic life, higher education — that is, the consolidation and onward transmission of established knowledge — is quite as important as producing more of it. The traditional combination of teaching, institutionally, departmentally and individually, with research is a major influence on the type of knowledge that is produced. On the one hand, it means that academic scientists are expected to produce findings that they can present coherently in lectures, articles and books. On the other hand, students are stimulated to question this putative knowledge, and thus to influence its form and substance. Discourse takes precedence over practice. So it follows that the library, rather than the laboratory, becomes the ultimate source of authority.

But the scientific archive is seldom encountered as a material entity. It is essentially a virtual institution located in a communication network. In principle, it encompasses the contents of all the periodicals, books and databases that are currently accepted as 'scientific'. As we have seen, research results are only officially recorded in this archive after systematic critical processing — 'peer review' — by expert referees.

What is not so widely recognised, however, is that acceptance by a peer-reviewed journal is only the first stage in a longer, less formal but more searching procedure. Once they are published, research claims are open to the whole academic armoury of sceptical scrutiny, theoretical counter-argument, experimental refutation, and so on.

---

[14] Ziman, J. M. (1978). *Reliable Knowledge: An Exploration of the Grounds for Belief in Science*. Cambridge, Cambridge UP, Ziman, J. M. (2000). *Real Science: What it is and what it means*. Cambridge, Cambridge UP.

[15] Ziman, J. M. (1968). *Public Knowledge: The Social Dimension of Science*. Cambridge, Cambridge UP.

Even formally codified scientific discoveries are not protected from further questioning. It is surprising how often a widely accepted scientific 'fact' — the immobility of the continents, for example — turns out to have been wrong.

That is why academic scientists refer to their knowledge as 'uncertain'. Of course they are sometimes tempted into describing it as absolutely 'true'. But this is not justifiable, for there is no procedure for closing off the critical process. The most that ought to said of a fact or theory that has survived the barrage is that it is now accepted 'objectively' by the relevant research community. Since nobody seems able to think of any further reasonable objections, working scientists simply become accustomed to using it without question in their own research. This epistemological uncertainty is well illustrated by the terminology of evolutionary biology. Scientists do not talk about Darwin's *Law* of Evolution by Natural Selection. So Creationists seize joyfully on their continued use of the phrase Darwin's *Theory*, as if there still lingered some doubt about it, long after it had become the cornerstone of the whole of modern biology.

We are told that 'Mode 2' knowledge production — in effect, technoscience — arises in 'contexts of application'. We might then say that 'Mode 1' — i.e., academic science — exists in a context of 'discovery-and-verification'. Philosophers used to make a distinction between these two components of the research process, but now recognise that they are inextricably entwined. Academic science is famously proud of its 'creativity': more soberly, it is obsessed with concerns about its own credibility. A great many of its social norms and epistemic practices are devoted to the validation, assessment, evaluation, testing, etc. — in a word, 'verification' — of its knowledge claims.

Take, for example, the 'experimental method'. This is not, as is often asserted, a defining feature of all scientific activity. Logically speaking, it can never *prove* a scientific proposition. But it is an extremely convincing procedure for replicating a discovery, putting a hypothesis through the mangle of practice,[16] disconfirming a conjecture,[17] selecting the fittest meme,[18] or whatever it is that scientists do to make sure that they are getting things more or less right. What more could we ask for, as mortal beings in a world we did not make ourselves?

Academic Science as a dynamic system

Surely, I must be joking! Everybody knows that academics are arrogant, aloof, dogmatic, quarrelsome, pedantic, vain, elitist (etc.), and essentially useless when it comes to practical matters. How can I paint them in such rosy colours? Have I just fallen for their propaganda? All that stuff about the 'scientific attitude' is ideological whitewash. Rub it off, and you will find ordinary people making up stories about themselves and each other in endless battles for social influence and ascendancy in science.

Well, that's true, just as it is of every vocational group — doctors, lawyers, politicians, priests, undertakers, and all. Underneath the snowy garments of an exalted calling, are human bodies with earthy thoughts and earthy needs. The value of much sociological research on science is that it has revealed the subtle ways in which the struggle to meet these needs is conducted and its effects on the form and substance of scientific knowledge.[19] But these studies systematically 'bracket out', on principle, the 'superstructure' of norms and conventions that define specific callings and differentiate them from each other. So having decided that physics professors must be considered on exactly the same terms as advertising executives — or, for that matter, Zulu warriors — they sagely discover that there is nothing to choose

[16] Pickering, A. (1995). *The Mangle of Practice: Time, Agency and Science.* Chicago IL, U of Chicago Press.

[17] Popper, K. R. (1963 / 1968). *Conjectures and Refutation: The Growth of Scientific Knowledge.* New York NY, Harper Torchbooks.

[18] Cziko, G. (1995). *Without Miracles: Universal Selection Theory and the Second Darwinian Revolution.* Cambridge MA, MIT Press.

[19] Latour, B. and S. Woolgar (1979). *Laboratory Life: The Social Construction of Scientific Facts.* London, Sage, Knorr-Cetina, K. D. (1981). *The Manufacture of Knowledge: An Essay on the Constructivist and Contextual Nature of Science.* Oxford, Pergamon.

between them from a moral point of view. Big deal — but only for those who have taken all the rhetoric about the 'scientific attitude' at its psychological face value.

A more fruitful sociological truism, however, is that various vocations — medicine, law, politics, the church, undertaking, advertising, etc. — perform various social functions. They necessarily differ greatly from one another in their practices and precepts. The most fascinating finding of the human sciences is the subtlety with which these diverse conventions and norms dovetail together into a more or less stable social order. Whatever they may be privately thinking, or covertly doing, academic scientists have to play by the rules of their profession. And there is ample evidence that they normally perform their prescribed public duties reasonably diligently. Indeed, the normative force of the academic ethos is demonstrated by the trouble that some of its deconstructors take in 'unmasking' deviations from it!

But this study of the place of science in society does not pretend to discover, display, disconfirm or disenchant any general sociological principles. I am just pointing out that many of the established traditions of academic science are conducive to its performance of several valuable non-instrumental social functions. One could rightly argue that these particular functions are evident to *us,* and seem indispensable, because we are in these actual historical circumstances. Our pluralistic social order and our pluralistic academic science evolved together, and are thus closely adapted to each other. Had other forms of society — e.g. Joseph Stalin's totalitarian technocracy — survived, they would no doubt have developed modes of knowledge production capable of fulfilling such other non-instrumental functions as would still have turned out to be necessary. But that would be in another country — and besides, that particular tyrant is dead.

What has to be admitted now, however, is that academic science is not nearly as well-fitted to these somewhat idealised functions as I have enthusiastically suggested. This is not just because it is a sub-optimal institution peopled by frail human beings. Several of its basic structural features are actually quite discordant with its exalted ethos.

The question to ask, for example, is how academic scientists are induced to produce knowledge in conformity to these unwritten principles. This is not because they are intimidated by negative sanctions. One of the surprising characteristics of established cultural systems is how feeble and unsystematic they often are in punishing infringements of their norms. Academic institutions are under pressure nowadays to set up more elaborate mechanisms for detecting and dealing with plagiarism, fraud and other gross infractions of their customary procedures. And yet, although the actual incidence of such deviance is a matter for conjecture, anecdotal evidence suggests that it is remarkably low by comparison with most other professions.

On the contrary, academic scientists are kept in line by carrots rather than by sticks. Good work is publicly recognised and rewarded: while bad work is simply ignored. At the lowest level, the carrot is only a poorly-paid, non-tenured job, such as a post-doctoral fellowship. Further up the ladder, it is a dignified permanent post on a comfortable income. At the highest level, however, the reward is not, as in other professions, big money and/or socio-political power: it is social esteem. Academic scientists, like everybody else, need to be paid, and seldom reject bureaucratic authority when it is offered to them. Indeed that 'recognition' by their colleagues is what they primarily strive for, and can never have too much of.

That sounds both high-minded and cynical. So let me put it round another way. For one reason or another, apprentices to research enter a profession where *credibility* is the token of personal quality. This is the password, the PIN number, the credit rating, for regular employment and for all the other tangible and intangible rewards of a scientific career. Without it, there is no future except as a hack teacher or technical assistant.

But scientific credibility is a fading asset which has to be continually renewed. What is more, it is an attribute visible only to other members of a research community and can only be acquired or refreshed by competent participation in their activities. This requires, above all, the contribution of acceptable research findings to the archives. To be acceptable, these findings and the way that they are presented must satisfy the established conventions of academic science, as assessed by the relevant community. Thus, academic scientists are strongly motivated to follow these conventions, as best they can, for the rest of their lives.

The place of science in society includes and depends on what it offers to individuals by way of a professional career. Attempts have been made to explicate this in quasi-economic terms,[20] but it is more authentically presented in fictional form.[21] The social psychology of academic life is evidently compelling, for it has operated effectively for the best part of two hundred years. But as was famously pointed out by Max Weber nearly a century ago,[22] it is not free from psychic traumas. For those of us who succeed in realising our youthful passions through it, it works like a dream. For less fortunate aspirants, on the other hand, it sometimes leads to a very disappointing and anxious life.

The competitive culture of high academic science is ultimately what makes it credible and reliable. But it is driven by personal anxiety. Even the most celebrated individuals are uneasily aware that both their achievements and their rewards are 'social constructs'. A scientific 'discovery', whatever its basis in material reality, has to be 'recognised' by the community to which it is reported. And even a prestigious prize, however gross in hard cash, is valued primarily as a symbol of that recognition. It is said that Niels Bohr, one of the most famous theoretical physicists of the twentieth century and one of the originators of the Quantum Theory

---

[20] Ziman, J. (2002). The microeconomics of academic science. *Science Bought and Sold: Essays in the Economics of Science.* P. Mirowski and E.-M. Sent. Chicago, University of Chicago Press: 318-40.
[21] Cooper, W. (1952). *The Struggles of Albert Woods.* London, Jonathan Cape.
[22] Weber, M. (1918 (1948)). Science as a vocation. From *Max Weber.* H. H. Gerth and C. W. Mills. London, Routledge & Kegan Paul: 129-56.

still remained doubtful in his old age whether his name would even be remembered in time to come!

To appreciate fully the force of this feeling one must grasp a paradox. Academic science, the epitome of a collective enterprise, is fuelled by possessive individualism. If we think of it as a magnificent building, like a mediaeval cathedral, then one of its most imperative traditions is that the names of the craftsmen should be clearly displayed on the elements of the structure, whether in the keystone of the great vaulted roof, the delicately carved ornamental bosses and gargoyles, or just the hidden blocks of masonry in the foundations. That is to say, every item of public knowledge accumulated in the archives is carefully labelled with its author or authors, whose names must be cited in any subsequent references. It is through this residual 'ownership' of their respective contributions to the communal storehouse, that academic scientists preserve and enhance the credibility that is so vital to their professional careers.

But as we have noted, credibility is an intangible attribute that can quickly decay or be debased. It is not surprising that academic scientists oscillate between boosting their own ideas by asserting them dogmatically, and putting out a smokescreen of agnosticism behind which, if necessary, they can retreat without serious loss of face. What is surprising though, and often very damaging to their social role, is the naïve self-confidence with which very successful academic scientists sometimes hold forth on subjects far away from their acknowledged areas of competence. It is quite embarrassing, for example, to listen to very eminent physicists discoursing on philosophy, distinguished biologists opinionating on social anthropology, or famous chemists telling the world what is wrong with religion, just as if these foreign fields had none of the competing paradigms, disputatious professors, and profound enigmas of knowledge which are so familiar on the territory of their home farm.

It is important for scientists to have a voice in society, but not as cock-sure know-it-alls. They are not like philosophers, with very wide interests — indeed with concerns that are so general that they finish up admitting that they 'know nothing about everything'. On the contrary, academic scientists are *specialists*, who get to know more and more about less and less until, it seems, they 'know everything about nothing', or nearly nothing! This sharpness of focus, this tunnel vision, is so notorious that it is often supposed to be an innate trait of character amongst scientists. But it is clearly what the French call a *déformation professionelle*. Although not usually included in any description of the academic ethos, it is an essential element of it.

The social psychology of scientific specialisation is very simple. Academic scientists are required to exercise their personal autonomy and demonstrate their originality by undertaking novel research projects. This is effectively impossible without a great deal of knowledge about what has previously been discovered or is already under investigation by others. The whole scientific archive is enormous. So the conditions for professional advancement can only be satisfied by narrowing down the search for do-able problems in a very restricted region—one shelf perhaps in the fabulously limitless Library of Babylon. By mastering and adding a few pages to the volumes on that shelf one can at least pursue an honourable career in ones chosen vocation. And the fact is that the great majority of research scientists follow that principle throughout their working lives.[23]

The trouble is, though, that what the scientist presents as her own peculiar obsession is not entirely of her own making. It is almost always an area of knowledge and research activity already recognised as distinctive by other scientists. People have probably been working in that neighbourhood for years. In the large standard academic terminology, it is one of many 'disciplines'. Universities list the hundreds of 'faculties' and 'departments' where these are taught as separate 'subjects'. Then the research world is even more finely subdivided, into 'sub-disciplines', 'fields', 'sub-fields', 'problem areas', etc.

---

[23]   Ziman, J. M. (1987). *Knowing Everything About Nothing: Specialization and Change in Scientific Careers.* Cambridge, Cambridge UP.

The differentiation of knowledge and its production into 'specialties' has important philosophical implications which are only just beginning to be explored.[24] It is also a matter of great sociological interest. What are we to make, for example, of the international 'invisible colleges'[25] where hundreds of specialists gather to discuss their common research problems — and seek employment, display personal credibility, manage communal resources, exercise social authority, and so on? The conventional academic ethos treats these simply as communication networks which facilitate, rather than restrict, the freedom of their members to operate as autonomous individuals. Nevertheless, they are powerful, long-lived social institutions, which have always been a feature of academic science and strongly shape its participants and their products.

One might think that undue specialisation is just an internal issue for the research system. But it also affects the place of science in society. In the first place, scientific experts are labelled for public use in terms of their specialties. However little we really know about Professor Bloggs or Dr Cloggs, we do have the assurance that they are both genuine experts on certain genera of veterinary gastro-intestinal flora, and thus fully competent to advise us on the production of methane by farm animals. That at least is something — although it doesn't give us the more relevant information that the Prof. is a clapped-out old hack who has long resisted all questions about climate warming, whilst the Doc has only just got his PhD for a molecular analysis of the segment of DNA that apparently protects these particular bacteria from the powerful acids in the stomachs of ruminants, and probably knows nothing else about agriculture. In other words, the ways in which academic scientists define their specialties bear little relation to the forms in which their expertise may be needed in the outside world.

---

[24] Ziman, J. M. (2000). *Real Science: What it is and what it means.* Cambridge, Cambridge UP.
[25] Price, D. J. d. S. (1963/1986). *Little Science, Big Science — and Beyond.* New York NY, Columbia UP.

The fact is that academic science does not map neatly on to the problem areas of the life-world. Its division into disciplines evolved as a scheme for classifying the research activities of scholars, not of the universe they were studying. It arose out of their competitive striving to demonstrate their individuality, not out of a collective endeavour to represent the state of things as a whole. This is not to say that its major categories — the Natural Sciences and the Humanities, Physics, Physiology and Phonetics, Microbial Genetics and Public-Sector Economics, etc. — are arbitrary constructs. They do have to be more or less consistent with the realities to which they refer. Very convincing reasons can usually be given for the criteria by which they are typically defined and differentiated. But this rationale is essentially historical. The academic 'world picture' purports to depict the way things are: many of its features — especially its subdivisions — are hangovers from the course that this knowledge has taken during its production.

Remember that academic scientists have to have some degrees of freedom in their choice of research problems. In exercising this freedom, they obviously have in mind the likely actions of their known 'peers'. But there is no overall authority, no general plan, directing these choices. So the outcome of all these individual, locally interactive decisions, however intelligible and reliable their findings are in detail, cannot be supposed to have any coherent meaning as a whole. In particular, there is no reason to believe that an 'invisible hand' will guide the growth of knowledge for the good of society. The mess of facts and theories, data and techniques, which have accumulated in the scientific archives can be abstracted, classified and searched for nuggets of enlightenment and gems of wisdom. But the notion of generalised beneficial 'progress' simply cannot be distilled out of the whole stewpot of ideas.

And even if we supposed that academic science would eventually generate a reliable, coherent, comprehensible map of 'everything', its advance in that direction is very unsteady. The way in which knowledge is produced makes it very subject to fads and fashions. Desperately trying to be at the forefront, researchers are strongly influenced by the plausible conjectures of charismatic individuals, who all try to 'jump on the bandwagon' setting off through the supposed 'breakthrough'. Conversely, a topic that is generally deemed not to be 'interesting' or 'exciting' is neglected by the overwhelming majority of the researchers competent to study it, thus leaving large gaps in the knowledge base. So academic expertise is very patchy. On some important subjects it scarcely exists at all; on others it is pedantically esoteric, confusingly controversial, or peculiarly self-deceived. Perhaps that is why it so often fails, in the end, to perform all the non-instrumental roles demanded of it by our rapidly changing society.

But Research Cultures Are Changing

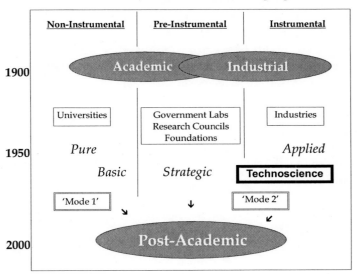

SEVEN

The Transition to Post-Academic Science

The conjoined twin traditions

Whenever I try to write about Academic Science I face a peculiar dilemma: should it be spoken of in the past or the present tense? Am I referring to a cultural form which is now an empty ritual, or to one that continues to be actively practised? Should I say, for example, that academic scientists *used* to have to publish all their findings, or that they *still* have to do so? Was most of my last chapter just a flowery epitaph for a noble but dead tradition; or does it describe a contemporary way of life? Young scientists fighting their way on to the faculty lists of research universities may scoff nowadays at the notion of 'getting tenure': 'that will only come when I'm beginning to think of my pension', they say, 'and anyway it will all depend on how much grant money I can bring in from industry'. Has that high-minded academic ethos gone for good, along with the practices that sustained it?

'Not yet,' is the short answer. A great many people, young and old, still largely follow its norms and conventions. Many of the institutions remain outwardly the same. The ivy-covered walls of the great universities—Oxford and Harvard, the Sorbonne in Paris and the Eidgenosse Technische Hochschule in Zurich—still shelter rivalrous dons in their hundreds and disrespectful students their thousands. It is still quite feasible for a clever young person to enter a university and climb the traditional scholarly ladder, from PhD to distinguished professor, solely on the basis of their original contributions to knowledge. The scientific archives are still mushrooming with just such contributions. Peer review is as demanding and contentious as ever it was. In some fields of study, such as pure mathematics, academic business goes on almost exactly as usual.

Nevertheless, behind this façade of normalcy, science has been going through an internal revolution. In less than a quarter of a century, the social machinery of the knowledge-production system has been subtly transformed. In particular, the *conduct* of research in institutions of higher education is very different nowadays from the *conduct* of 'academic science' of the past. The same words are used to designate its practices, offices, duties, privileges, rights, responsibilities, etc., but they no longer mean the same as they did fifty or a hundred years ago.

The difficulty in perceiving this historical trend is compounded by the fact that even in its heyday academic science was not a standardised social form, sharply differentiated from other ways of knowing or getting to know.[1] The account presented in the previous chapter was an idealised stereotype whose features merely typified a great diversity of national practices. It also put the whole spectrum of academic disciplines on the same wavelength. It cannot be that physics and philology, astronomy and agronomy, economics and engineering, epidemiology and education, were all alike in the way that they encouraged, assessed and rewarded original contributions to public knowledge.

Even in the area of our main concern — the distinction between instrumental and non-instrumental science — there was never a sharp line of demarcation between them. Broadly speaking, the institutions devoted to the academic mode of knowledge production did not go in for the type of

[1]   Pickstone, J. (2000). *Ways of Knowing: A New History of Science, Technology and Medicine*. Manchester, Manchester University Press.

research and development undertaken in industrial laboratories, and *vice versa*. Individual scientists mostly followed
one or other of these traditions throughout their working
lives. But people could more easily cross the interface some
decades back, and very often did so. For example, industrial
firms were always on the lookout for young academic
scientists already trained in advanced research, and
they sometimes appointed well-established university
professors to head their R&D divisions.

Knowledge, also, could flow without formal barriers
between the two research cultures. Or, rather, it could flow
freely 'down' from academia, where it originated, into
industry, where it was put into practice. This so-called
'linear model of technological innovation' is the great foundation myth of the twentieth-century research system.
Great ideas, so it held, spring out of soil that has been deeply
cultivated by academic scientists. But these ideas are materially powerless until they have been adopted and nursed to
their full strength in the busy workshops of trade and
industry. So academic research lacked direct instrumental
capability, but it was always valued and exploited for its
pre-instrumental products.

As we have seen, that model grossly underplays all the
knowledge that is actively produced in the invention,
development, design and manufacture of technological
artefacts. It also neglects the feedback from technoscientific
practice 'up' the information channels into the academic
sphere. The main point is that knowledge has always
flowed across the interface between these two research
cultures, and is relied upon as 'scientific', no matter which
side it originally came from.

Indeed, the signal virtue of the European university and
its global progeny is that it has always been open to the
practical arts. Medicine and Law, for example, were
amongst its founding faculties. In modern times it has
added Engineering, Architecture and Business Administration, etc., not to mention a variety of less exalted but equally
useful professional curricula such as Pharmacy and Media
Studies. The main social function of higher education is the

training of practitioners in conceptually sophisticated callings, which is much better done in an active research environment. That demands, in turn, a determined effort to distill a more or less 'scientific' discipline out of a heterogeneous collection of tacit skills and empirical techniques. Thus, to teach a subject such as forestry, town planning or clinical medicine requires a grasp of both theory and practice. Departments, faculties — even whole institutions — that offer higher education in such subjects have always had to straddle the boundary between the two research cultures.

At the same time, the emergence of new 'science-based' technologies created further linkages. The electrical and chemical industries, for example, developed their own internal technoscientific environments. But they took care not to cut themselves off entirely from their roots in academic science. Some of the largest firms established research laboratories where scientists could work on problems of their own choosing, while figuring at the same time in scientific circles almost as if they were university professors.

The fact is that in the tug of esteem between these two modes of knowledge production it was usually the academic side that pulled the hardest. Until recently, anyway, its ethos has normally prevailed. Even in a newly-created 'Institute of Technology' — in Massachusetts, California, Manchester, Gothenburg, Delhi, Tokyo, wherever — research activity followed university conventions. No doubt many senior members of staff were spending a lot of their time — and earning a lot of their money — as expert consultants in the outside world. But this was considered their own private business, of no more significance from an institutional point of view than if they were Members of Parliament or Masters of Fox Hounds. And as with the staff of teaching hospitals, bureaux of standards, agricultural field stations, etc. many of their research projects were formulated 'in contexts of application'. But in the end their findings had to be presented as public contributions to knowledge, entirely disconnected from their instrumental background — and especially from sordid monetary motives!

The two modes of knowledge production were thus like conjoined twins, separate in heart and mind but intimately connected in body and spirit. The public saw them as essentially a single organism, but they played different roles, and did not compete directly for esteem or resources. The technoscience twin, quietly making money in a hidden corner of society, was perhaps in some ways the more influential. On the other hand, its academic twin, winning esteem for its occasional virtuoso public appearances, was deemed to be dominant. So people enjoyed all the technological goodies of 'progress' — the railways, automobiles, telephones, surgical anaesthetics, refrigerated food, high explosives, etc. — but attributed them to the amazing discoveries made in the search for 'truth'. Thus nurtured and protected, academic science was enabled to perform its non-instrumental functions without serious challenge.

The collectivisation of Academic Science

Academic science began to take on its modern form around the beginning of the nineteenth century. Technoscience — the industrialisation of invention — appeared in Europe and North America towards the end of the same century. The two modes of knowledge production remained distinct, although always in close association, for another hundred years. During that time, however, they were both changing, in their different ways.

Technoscience, as we have seen, grew from modest beginnings in the small back rooms of industry until it had become the driving motor of the economy. By the end of the Second World War (which it had only just managed to win for one side and only just lost for the other), it had developed the institutional characteristics discussed in an earlier chapter. 'Research Management', or 'the Administration of Research and Development' had become a distinct profession.

This managerial culture developed primarily to cope with an enormous expansion in scale and scope. From its traditional base camps in electrical, mechanical and chemical engineering, armament manufacture, hospital medicine, etc., technoscience had invaded the whole economy. It was everywhere: in the private and public sectors, in the home and the marketplace, in the countryside and in the city, on sea and land and in the air, in the arts of peace and in the weaponry of war. To determine its 'place in society' one just had to look around!

As a natural consequence, it began to register officially as a major factor in the economic order. Instead of trying to make out whether a firm was winning customers or making profits, stock market analysts looked at its investment in 'R&D'. Instead of asking whether a government agency was actually performing its functions efficiently, effectively or efficaciously, legislators questioned whether it was using the very latest technology. To decide whether a country was economically 'developed', one only had to check that it had a high ratio of R&D to GDP—i.e. checking that its aggregated expenditure on the numerous varieties of technoscience amounted to several per cent of its national income. 'Science Policy' (strictly speaking, technoscience policy) had become a recognised branch of socio-economic discourse.

During this period, academic science also boomed. From its furthest origins, back in the seventeenth century, the production of scientific knowledge had always been a growth industry.[2] Measured in its own standard units—total number of publications—it was just going on doubling every fifteen years. But from the early nineteenth century onwards, its social base and working space in higher education was also expanding rapidly. Technoscience and science-based professional practice promised employment to science graduates. Growing national populations, enriched by the Industrial Revolution, were sending a higher proportion of their young people to the many newly-founded universities. Academic science was spreading into old civilizations such as India and Japan and into newly colonised territories

[2]   Price, D. J. d. S. (1963/1986). *Little Science, Big Science — and Beyond*. New York NY, Columbia UP.

such as the silicon valleys of the American West. Traditional disciplines were splitting and branching, and whole new research fields were being discovered or invented — biochemistry, social psychology, archaeology, and many others. The Sciences, old and new, were where the action was, and from there they were attracting more and more students from the old territories of the Arts and Humanities.

The key point is that scientific research established itself professionally in academia in the second half of the nineteenth century. Thereafter it grew enormously in scale and scope for more than a hundred years without coming under significant pressure to change its customary practices and ethos. If one of the old Ivory Towers was getting a bit crowded, there were always open spaces where enterprising fledglings could build themselves new ones. Competition for the top posts was always fierce, and the lower ranks were often very poorly paid. But there were several strata of less prestigious universities, colleges and schools where a disappointed high-flyer could find a decent job and still have some time and facilities for scholarly work. So personal disaffections did not clump together into a movement for internal reform.

At the same time, society was quite reasonably satisfied with what it was getting out of this peculiar institution. Academic scientists were making amazing discoveries, from which, in the end, great benefits would be bound to flow. The way that they organised themselves might seem extremely laborious and inefficient, and much too tolerant of sloth. But their reputation for sincerity and probity was unequalled. They were all committed personally to the production of knowledge; and enough really good research was being done by the outstanding members of their community to cover the very modest achievements of the majority. Anyway, so long as they all performed their teaching duties reasonably conscientiously, why worry?

Furthermore, there was little public interest in the research agenda. From a present-day standpoint, it seems remarkable that academic scientists were not being urged to tackle practical issues, apply their research to national needs, solve outstanding societal problems, and so on. Of course there was a warm welcome for potentially useful discoveries, whether in the form of artefacts, techniques or concepts. But these were supposed to come incidentally out of the search for understanding. It was a matter of principle that academic research could never be planned. The deliberate production of instrumental knowledge was the responsibility of technoscience, over on the smoky side of town. And as we have seen, academic science has a natural facility for spinning a dozen new questions out of the answer to every old one, so there was never a shortage of exciting problems to be attacked. The frontier of knowledge was not only endless:[3] it lengthened as it moved forward.

This idyllic state (from the point of view of the scientific Establishment) lasted for a remarkably long time. But it had one serious weakness: research is not a cost-free activity. What was worse, it becomes steadily more costly. Intellectual progress is always opening up new problem areas. Technological progress makes it feasible to advance into them. But sophisticated research equipment is expensive to buy, house, and service. So the sheer cost of producing a new item of knowledge — a scientific paper, for example — increased exponentially, decade by decade.

Technoscience can always justify its rapidly expanding budgets in terms of future profits. It may have started as a speculative investment, but in the long run it often paid off very handsomely. But academic scientists can only hint vaguely at the potential benefits, like the future usefulness of new-born babies (see Faraday's throw-away comment of the previous century). As we have seen, financial outcomes depend more on financially benevolent patrons than on the fortune-seeking investors. At first the costs of research were a relatively modest addition to the ordinary scholarly stipends of the university involved, and could be met from normal monitory sources, such as student fees and endow-

---

[3]   Bush, V. (1945). *Science — The Endless Frontier: A Report to the President on a Program for Postwar Scientific Research*. Washington DC, National Science Foundation.

ment income. But as time went on, more and more money for academic research had to be sought actively from outside patrons, such as private foundations and national governments.

Indeed, by the beginning of the Second World War, academic science in most countries was already heavily dependent on state funding. This came in various forms. Some of it was included notionally in the block grants or regular budgetary subventions to the universities. Some was channelled into specific subjects, such as medicine, through quasi-autonomous non-governmental research councils. Postgraduate studentships provided a meagre living for useful apprentice researchers aspiring to doctorates. In some European countries, state-funded organisations were set up to support well-qualified research scientists for whom there were no regular university posts. In other words, even in the depths of political turmoil and economic depression, academic science was publicly acknowledged as a national asset, which no self-respecting country should allow to fall into decay or disrepute.

Nevertheless, academic research was still a select profession, manned largely by a few men (and a mere handful of women) with a yen for it. This marginal stance changed radically in the 1940s. As the phrase goes, academic science 'had a good war'. The professors and lecturers were put to work, sometimes in their own laboratories, on the invention and development of novel weapons or other technological products of military value. As we all know, they succeeded magnificently. This was absolutely instrumental science, of course, but even the most highbrow academics in apparently irrelevant subjects proved to be rather good, also, at technoscience. So when most of them returned to their universities after the War, they sought, and received, their due reward. That is to say, very large funds began to flow from the public purse into academic research.

But there was a problem. How was this expenditure to be justified? It was all very well for millions, billions, to go to big technoscience projects like nuclear power systems and space satellites. These could be defined as instrumental to the prosperity and security of the nation. The covert militarism of the Cold War sheltered many technological extravagances. But why should good public money go to, say, a reclusive neurophysiologist studying what happened in the brains of cats when they saw dogs, or to a compulsive physicist who had a scheme for following Einstein's ideas for the observation of hitherto invisible gravity waves? Investigative journalists and adversarial politicians were adept at uncovering such 'wasteful practices' and sometimes recruited members of the public to join in their noisy protests. In the large, academic science could always claim a strategic role. But its academic members found this contrary to the whole spirit of science, as they perceived it, to have to provide a specific instrumental rationale for every last, loneliest and most exquisite project on its research agenda.

The trick was to launder the money through a series of 'buffering' organisations, such as research councils and universities, where funds could be allocated to the actual sites of the research by innumerable panels of specialists. Typically, these panels and their secretariats would take the peer review of publications as their model. Thus, they would receive project proposals from individual scientists, send these proposals out to expert referees for critical appraisal, and then decide whether to satisfy their requests for grants. The amounts to be awarded by the various panels would be decided by a hierarchy of committees, manned similarly by academic worthies, reaching up to the top of the organisation. In this way, the funding of academic science was put into the hands of the academic community, and were not be questioned by any lay person.

So this system, or something rather like it, gave academic science the material means to go on expanding at the traditional rate without any radical change in its practices and ethos. Indeed, wartime experience of exceptionally well-funded technoscience encouraged the academic scientists to demand bigger and better research facilities. At first their requirements were relatively modest. A million dollars, say, for a high-voltage electron microscope, together with the salary of an exceptionally skilled technician to run it.

That might involve a bit of work behind the scenes to get sufficient support from potential referees, but would not break into the annual budget of the grants panel.

The trouble is that research apparatus operates at the frontier of technical sophistication. After a couple of years or so, our beautiful instrument is already looking a bit out of date. We yearn for the new model (recently installed by a rival research group!), which is ten times as powerful, productive, stable, discriminating, efficient, etc. — and only twice as expensive. And as these magnificent machines take up valuable laboratory space, we mustn't forget to indent for the buildings needed to house them, and the additional computer facilities they will employ. Furthermore, they are so productive that we ought to ask for stipends for more PhD students and postdoctoral research assistants to spin really good science out of them.

In some fields, indeed, the apparatus required to do competitive research soon became enormously elaborate and expensive. By the late 1970s, a good experiment in high energy particle physics could only be done by a team of hundreds of fully qualified scientists and engineers working together for years on a 'facility' costing of a billion or so dollars to build, and hundreds of millions annually to run.[4] Needless to say, the enormous sums required to fund this type of 'Big Science' didn't flow out of various public purses without months of committee work, acres of typescript, and persistent, systematic, political lobbying.[5] Each national government, each official culture, each international organisation, developed its own Byzantine procedures for financing its 'science base'. And yet, although the knowledge thus produced was usually highly 'academic', it continued to be resourced with remarkable generosity and even indulgence, as perceived from the outside.

---

[4]   Ziman, J. M. (1976). *The Force of Knowledge.* Cambridge, Cambridge UP.

[5]   Greenberg, D. S. (1969). *The Politics of American Science.* Harmondsworth, Penguin.

It is often said that a 'contract' had been established between science and society.[6] But this contract was never written down in full. It was simply implied by a raft of elaborate political procedures whose small print was indecipherable. The opaqueness of these served to keep the relationship as loose as possible. The over-riding principle was that nothing should constrain a handsome flow of money in exchange for a correspondingly fruitful flow of discoveries. To politicians and public alike, science was a fairy-tale goose. Stuff its mouth with gold, but give it a free range in which to lay its magic eggs.

In effect, academic scientists could still obtain the means to tackle research problems of their own choosing. It was not just that they were allowed to exercise freely their technical and methodological skills. What was also being enabled at the same time was the *strategic autonomy* which is so central to their ethos. So they could go on competing for credibility and recognition by creating, publishing and criticising their knowledge claims, just as they always had done. It seemed that society had provided academic science with a much-deserved paradise!

True, a project could only be undertaken if one could persuade ones fellow scientists that it was worth attempting and could reasonably be carried out by the proposed methodology. But this was well within the spirit of academic science. After all, one could not publish ones cherished findings unless ones 'peers' thought them credible. Researchers who were adept at dealing with referees soon added 'grantmanship' to their social and rhetorical skills. The funding bodies and their subject panels were familiar colleagues wearing other hats. And when it came to the planning of really big projects, the formal and informal networks of ones research community could easily be activated. Indeed, one of the remarkable features of this period was the emergence of enormous 'collaborations' in which

---

[6] Bush, V. (1945). *Science – The Endless Frontier: A Report to the President on a Program for Postwar Scientific Research*. Washington DC, National Science Foundation.

literally hundreds of independent academic researchers, located organisationally in many different institutions — in some cases in a different countries — worked voluntarily together on an immense scientific experiment with the bare minimum of top-down managerial supervision.

So academic science adapted without a hitch to its enriched financial environment. Nevertheless, in order to exploit these more ample circumstances, it had had to moderate its traditional individualism. What this meant in practice was that certain vital personal decisions — what problems shall I now tackle, what apparatus and technical assistance should I seek, with whom will I try to work, etc. — could no longer be determined in accordance with their own idiosyncratic preferences.[7] The available options were strongly limited by the views of other people — members of grant panels, reviewers, funding body officials, potential collaborators, and so on. These constraints were softened by the fact that almost all of these other people were scientists like oneself, with a thoroughly expert understanding of the situation. Discourse with them, whether by persuasion, instruction or negotiation, could be conducted in the technical language of the subject. A grant application for a research project, for example, could be written in the same rebarbative style as the paper that would later report its findings. Sometimes — by a method that no non-scientist could fathom — the applicant might already have private evidence strongly favouring these findings, so the promised outcome was reasonably well assured in advance!

In effect, the weight of the scientific culture had shifted towards its communal foot. Seen from the outside, academic science was always much more of a social enterprise than it seemed to insiders. Few researchers could ever really have been described as 'lonely seekers after truth', 'solitary explorers of the natural world', 'private thinkers of the unthinkable', etc. They had almost always built on the work of their past and present peers, often as presented formally

---

[7]   Ziman, J. M. (1981). 'What are the options? Social determinants of personal research plans.' *Minerva* 19: 1-42.

in the literature of their subject. Now they had to reveal to each other — much more explicitly than they might ever have expressed them to themselves by choice — their plans for future research, and conform their 'projects' to more widely shared perspectives. In other words, the possible implications of their research, whether purely 'scientific' or potentially practical, had to be envisaged and offered for scrutiny.[8]

Furthermore, individual researchers were finding it much more difficult to go it alone. Even for quite modest projects, reliable knowledge could not be produced without what Soviet military strategists used to call a 'correlation of forces'. Scientific progress revealed problems that could well be solved by the application of a diversity of technical skills. But no single, narrowly specialised academic scientist could know enough about the instrumental, observational, theoretical and computational methods of several different scientific disciplines. It was not just a matter of building up a big 'research group' of post-doctoral assistants, doctoral students and 'supporting staff': it was necessary to team up, on an even footing, with other established experts in other fields, sharing the responsibilities and rewards of a joint contribution to the literature. The growing proportion of multi-authored papers is strong evidence of this trend.[9]

In a word, academic science became more of a 'collective' than a 'communal' endeavour.[10] It is true that even the largest 'collaborations' were voluntary associations of independent individuals, and small interdisciplinary research teams were seldom permanent. Tenured research scientists, in universities or other 'academic' institutions such as national research councils, could still usually obtain the resources they needed to do 'their own work', as they touch-

---

[8]   Weinberg, A. M. (1962). 'Criteria for scientific choice.' *Ibid.* 1(2): 158-71.

[9]   Price, D. J. d. S. (1963/1986). *Little Science, Big Science — and Beyond.* New York NY, Columbia UP.

[10] Ziman, J. M. (1983). 'The collectivisation of science.' *Proc. Roy. Soc. B* 219: 1-19, Ziman, J. M. (1995). *Of One Mind: The Collectivization of Science.* Woodbury NY, AIP Press.

ingly called it. But the whole scientific community, world-wide, was now criss-crossed and held together by a mesh of formal and informal agreements, contractual obligations and other connections between its members. In particular, a growing proportion of fully qualified professional researchers were lingering into middle age in temporary posts, funded by the short-term research grants that fuelled the whole system.

Nevertheless, quite remarkably, the process of 'collectiv-isation' was not carried through to its logical conclusion. By, say, the end of the 1970s, academic science throughout the economically developed world was mainly funded by the state. And yet it was not assimilated into the state machine. In many countries, it was true that all the researchers were officially state functionaries, subject to the duties and privi-leges of normal civil servants. But they usually enjoyed quite as much 'academic freedom' as the faculty members of independent universities in other countries such as Britain. Direct governmental influences on their scientific work were laundered out of their research applications by quasi-independent agencies, such as 'councils', 'founda-tions', 'centres', 'academies', etc. Even in the old Soviet Union and its satellites, academic science was largely a self-contained sector of the *Apparat*. The Lysenko affair showed, by scandalous exception, that it was still expected to man-age its own affairs in its own way under the influence of its own, scientifically proven, leaders.

This is an important point, for it contrasted strongly with the way that technoscience was organised. As we have seen, this was typically bureaucratic. The scientific activities of a large industrial firm, for example, would be concentrated in the box labelled 'R&D division' on the organisational chart. This box would be shown as subservient to higher levels of authority in the managerial hierarchy, and internally subdi-vided right down to the research group level. The whole structure was designed to serve the interests and obey the commands of the Board. At least, that was the principle on which such systems were constructed, although whether or not they ever really worked like that is quite another matter.

The achievements of industrial and governmental technoscience in and after World War II thus provided an influential model for the large scale organisation of research funded from a single source. The fact that state-funded academic science was not 'reconstructed' on exactly the same lines is highly significant. True enough, there were excellent arguments, typically presented by Michael Polanyi,[11] favouring the traditional 'academic marketplace' mode for the production of knowledge, especially knowledge that is so 'new born' that its future prospects are entirely obscure. These arguments have been deployed regularly by scientific leaders and have never been seriously refuted.

On the other hand, there were cogent reasons, famously articulated just before the War by J. D. Bernal,[12] for planning and directing science to serve society much more closely. This reasoning again, was continually rerun by political and economic leaders. Their concerns for the welfare of their nations and/or the efficacious use of public money were perfectly genuine. So they were fully within their rights to question the apparently 'irrelevant' and 'wasteful' practices of academic science. Its survival as a highly subsidised but politically independent institution was more a sign of the continuously adaptive *pluralism* of our culture, than of any rational social design. In effect, so long as the egg-heads were producing an invaluable flow of pre-instrumental knowledge they were worth the relatively modest cost of keeping them happy in their work.

Fortunately, this studied indifference to the inner workings of academic science meant that it was still able to perform the non-instrumental social functions that I have discussed in previous chapters. These functions, however, were seldom formally defined. The most that might be said for funding the more obviously 'useless' branches of science, such as pure mathematics, cosmology or mediaeval history, was that they had very high 'cultural' value. So it

---

[11] Polanyi, M. (1962). 'The republic of science: its political and economic theory.' *Minerva* 1(1): 54-73.

[12] Bernal, J. D. (1939). *The Social Function of Science*. London, Routledge.

was a sign of grace for a civilised country to subsidise them, along with the performance of operas and the preservation of ancient monuments. Otherwise, all those academic books and journals were to be considered incidental by-products of the serious work of training the qualified techno-scientists, engineers, doctors, architects, accountants, financial analysts, managers, planning officers, *et al.* required by an economically competitive nation.

Here we should note that there were some hidden clauses in that unwritten contract between science and society. If these were ever spelled out, which they were not, they might have stipulated that the academic system should keep itself aloof from the everyday life of the nation. Like an Established Church, it had to be careful not to meddle in the affairs of the State even though some high-minded clerics were allowed to express their high-minded thoughts. Relationships between them should always be conducted through the official channels, and at arms length. It was not necessary for ordinary people to understand science: that was solely the responsibility of professional experts. Close collaboration between the 'laity' and the 'clergy' — for example in reports on social issues, technological risks, etc. — was not to be encouraged, for this would confuse the vital distinction between technical facts and human values. Increasingly however these distinctions have grown weaker. It might be felt that the academic scientists, especially those in the human sciences, should not be drawn by their curiosity (and/or humanity) into the complex depths of government or business, for that might tempt them to give apparently 'scientific' support to some highly contentious political causes. And these were increasingly to be found littering the ground where only scientific facts might have been expected.

Thus, for example, one of the effects of the collectivisation of academic science was substantially to reinforce its norm of 'disinterestedness'. This undoubtedly facilitated its non-instrumental role as a source of independent, impartial knowledge and expertise which would not bring any extra money or fame to any particular scientists. But it did enlarge the gulf between science and the non-official elements of the social order just at the time when technoscience seemed to be strengthening its instrumental hold over the whole culture. The radical scientists who pioneered the establishment of *'Science Shops'* to bridge this gulf were responding, in good Marxist fashion, to a significant change in the socio-economic relations of their culture!

Do I exaggerate? I can't quite quote chapter and verse for this analysis. Scholarly interest has been focussed on the growth of 'science policy' during this period. Heaven forbid that I should try to rehearse all the convoluted administrative systems designed by anxious governments to induce the scientific elite to control their burgeoning budgets and to devote them, somehow, to the good of the nation. But there has been almost no formal study of the changes that were occurring at this time in the detailed structure of the 'research system', nor of the effects of these changes on its relationships with society at large. To my mind, the interdisciplinary discipline of 'science and technology studies'[13] is fixated on the 'external sociology' of science, whilst largely neglecting closely connected features of its 'internal sociology'. That is why I have given a lot of space in this book to matters that might seem only of interest to career-conscious professionals. I don't believe it's possible to understand the changing place of academic science in society without an explanation of how it actually, or supposedly, works, or used to work, in practice.

Science in a steady state

Some time around 1980, academic science began to lose its privileged place in the polity. There was no sudden fall from grace and there were no dramatic signs of a new structural transformation that was already taking place. There was nothing like a signed missive from a collection of outraged Nobel Laureates as they tried to storm the National Science Foundation, neither were there Ministers of Science

---

[13]   Jasanoff, S., G. E. Markle, et al., Eds. (1995). *Handbook of Science and Technology Studies*. Thousand Oaks CA, Sage.

who felt so humiliated by their budgetary constraints that they considered turning down the Prime Minister's call to office. Even the official execution of the project to build a Superconducting Super Collider by the US Congress in 1993 was treated as no more than a temporary career set-back rather than as the tragic end of one of the public heroes of our time. Nevertheless, by the end of the '90s, it was clear that just such a collection of minor revolutions had indeed occurred.

Like climate change, this revolution started long ago, and is still going on. Most scientists don't want to hear about it. They want to go on reading from the same scripts, and complain bitterly about numerous changes in the stage instructions — for that read *'proposal forms for research find-ing'*. But many continue to deny that the scenery has been shifted and that they are acting out a different play in a completely different theatre.

If you were to ask them to make a thought experiment of the following kind it might gradually become clear that the changes had indeed happened. Imagine that you have pro-grammed your computer to compile a Grand Concordance of present-day texts about the scientific life. It will contain a number of entries such as:

> accountability; allocation; appraisal; career; centre; competition; contract; efficiency; entrepreneur; evaluation; excellence; exploitability; facility, research; foresight; indicator; innovation; input; interdisciplinarity; management; output; overheads; patent; performance; priority; program; property, plurality: intellectual; quality; system, research; resources; selectivity; research, strategic; tenure, abolition of; training, research.

There is no need to explicate these words. Some of them obviously relate mainly to technoscience. But everyone in academia is wearily familiar with them. Many of them come trailing clouds of red tape and recent disappointments. They triangulate the institutional arenas where the research game is already about to be played.

Now go back thirty years. Think of carrying out a similar analysis on the texts of those days. How few of those words would have appeared in discourse about science in the 1970s. How few of the policies and procedures they refer to would have been practised — would even have been conceivable — at that time. That is a measure of the extent to which the whole culture of academic science had been transformed since the days when, in 1970, I first began to study the social relations of science under the title of a talk given on BBC Radio 3 called 'Science is Social', which outraged no one at all.

Challenged thus, experienced scientists agree that there have indeed been many major changes in their working environment. Furthermore, these changes do not seem to be just temporary glitches, or little local difficulties. Taken together they pervade the whole enterprise, from top to bottom, from vice-presidents for research to graduate students, from Harvard to Podunk State College, from cancer research to cosmology, from Alaska to Beijing. But it must be admitted that even those who have adapted successfully to the demands of the new mega working environment, are still reluctant to admit that they have been going through an irreversible transition to a new regime.[14]

What, in fact, had happened? Sometimes the complaint was that academic science has been grossly *bureaucratised*. But as we have seen, this is only partly true. The self-image of the research community as an autonomous 'Republic of Science'[15] is far from dead. The increasing weight of formal managerial practices is only one symptom of a deeper cultural change.

A more radical interpretation of the change in academic science is that it has been *industrialised*.[16] As we shall see, this also is only partly true. Corporate industry is much more

[14]   Cozzens, S. E., P. Healey, et al., Eds. (1990). *The Research System in Transition*. Dordrecht, Kluwer.
[15]   Polanyi, M. (1962). 'The republic of science: its political and economic theory.' *Minerva* 1(1): 54-73.
[16]   Ravetz, J. R. (1971). *Scientific Knowledge and its Social Problems*. Oxford, Clarendon Press.

influential on the campus than it used to be. But the production of academic knowledge has long been an 'industry' in its own right, albeit not managed along conventional commercial lines. Institutionally speaking, it is still distinct from organised technoscience which is fed directly into the market economy.

The notorious tie-in of science with the military/industrial complex suggests that it is being *militarised*. This is far from paranoic. A large proportion of contemporary technoscience is now undertaken for direct military application.[17] But the association of advanced science with war is a very old story.[18] Even the hero-figure of Galileo himself offered his best hand-made telescope to the doge and Senate of Venice as a weapon of warfare on the grounds that it would *'allow us at sea to discover at a much greater distance than usual, the hulls and sails of the enemy so that for two hours and more, we can detect him before he detects us'*. The moral implications of this connection are very serious but they lie outside the frame of this book. Here we treat it simply as typical of the way that knowledge is produced to serve the requirements of the predominant culture of its time.

A more mundane analysis[19] might show that academic science was beginning to exceed its credit limit, and society felt bound to call in its loan and realise its investment. Consider the open-ended, 'hands off' mode of collectivisation described above. An essential condition for this to operate was that sufficient public funds should go into academic science to meet all its reasonable needs. But the internal dynamism of the research system is irrepressible. There was never really any hope of expanding state patronage at a rate that would match it. It is only necessary to look at the numbers. As we have seen, academic research is a growth

[17] CSS (1986). *UK Military R&D: Report of a Working Party of the Council for Science and Society.* Oxford, Oxford University Press.
Salomon, J.-J. (2001). *Le scientifique et le guerrier.* Paris, Belin.
[18] McNeill, W. H. (1982). *The Pursuit of Power.* Chicago IL, University of Chicago Press.
[19] Ziman, J. M. (1994). *Prometheus Bound: Science in a Dynamic Steady State.* Cambridge, Cambridge UP.

industry whose indicators — personnel, papers, project grants, etc. — have to go on doubling every 15 years. No national economy could possibly sustain such a rate of growth for more than a few decades. Support for science was bound to take an increasing proportion of the discretionary funds of the state. Eventually it had to reach its 'limits for growth'.

So the 'contract with science' could only be a temporary arrangement. Quite soon, it would have to be renegotiated. In the 1970s, two factors combined to level off the state funding of science in many developed countries.[20] One was simply a general economic slowdown. Finance ministers could no longer afford to subsidise academic research with their customary enlightened generosity. But the other factor was more permanent. In effect, the total 'science vote' was now large enough to be politically visible. As it approached, say, one per cent of the GDP, it would even show up on a one-page summary of the national budget. Thus, it invited comparison with the sums devoted to more salient national needs, such as the police, the railways, the health service or the two new aircraft carriers demanded by the Navy. I am not saying that this level of expenditure on science was unjustifiable. I am simply pointing out that it could not avoid coming into direct fiscal competition with more deeply entrenched political constituencies.

This funding crisis, then, was real. A transition from expansive to 'steady state' conditions was to be expected.[21] But when it happened, it had unforeseen effects on the system. It wasn't just that the pace of discovery had been slowed down. Every traditional practice and norm was put under pressure.

Continuous expansion is a structural principle of academic science. Its enormous intellectual productivity requires it to make itself anew every twenty years or so. Its practices are based upon the assumption that there will be

[20] Ziman, J. M. (1978). 'Bounded science: the prospect of a steady state.' *Minerva* 16: 327-39.
[21] Price, D. J. d. S. (1963/1986). *Little Science, Big Science — and Beyond.* New York NY, Columbia UP.

need for two research jobs soon where there was one job before, that old subjects are never easy to kill off to make room for new ones, that almost all alpha-rated projects will eventually get funded, and so on. Putting a boundary round its inputs didn't just stabilise its output. It dislocated the machinery inside the famous 'black box' of governance. It disrupted an established web of intellectual debits and credits, and also the more uncomfortable balance of social obligations and expectations, moral rights and responsibilities.

Take, for example, career prospects. These were geared to the presumption that at least two of the best students of an established professor would carry on the good work by winning tenured posts for themselves. But in a static profession, only one such position would be likely to fall vacant at the right moment. So highly qualified researchers would have to stay for years in a series of short term, subordinate 'contract' posts, before gaining the right to operate as 'principal investigators' on their own account.

This, in turn, sharpened the competition for all forms of research employment. Appointment committees were impelled to be much more fussy about their procedures. Elaborate ranking criteria and arithmetical formulae were invented to make their decisions seem more acceptable. Numerical data were attached to such ordinary academic practices as publishing ones work, citing ones sources and getting cited in the literature. Journals were to be measured by their 'impact factors' and university departments by their 'research performance indicators'. This fashionable, pseudo-scientific managerialism was sustained by a neurotic desire to adapt conscientiously to 'steady state' conditions. But nobody really believes that it does any better in this way of working than by using the traditional mechanisms, that were sloppy enough to accommodate the important innate uncertainties of all research activity.

At the same time, questions began to be asked about the contribution of academic science to national life. Why weren't its projects more 'relevant' to the perceived political, economic, environmental, social, medical, technological (etc., etc.) needs of society? Shouldn't it be made more directly answerable for the way it was spending the taxpayers' money?

This demand for greater 'accountability' went much further than requiring detailed budgeting of every step in every research project, or challenging the 'overhead' costs that institutions sought to load on them. Governments started to take an active hand in the shaping of these institutions, and trying to maintain their 'excellence' by differential subsidy. The fate of certain Big Science projects was debated in legislatures, and the international standing of the national 'science base' became a matter for public concern.[22] Science policy even began to be noticed in the quality press.

More significantly, these demands were becoming more explicitly instrumental. Grant applicants now had to explain how their putative discoveries would ultimately benefit the citizen in the street. Peer review panels had to score project proposals in terms of their 'social merit' as well as their 'scientific promise'. Priority was given to specific socio-economic needs, and funds for academic science were channelled into programmes designed to meet them. And so on.

It was believed by many involved in the 'linear model' for technological innovation that it was working much too slowly, and ought to be hurried along. So academic scientists were instructed to liaise more closely with industry. They had to make sure that their discoveries were patented so that they could be fully 'exploited' — even before they were made. They were encouraged to parlay their 'intellectual property' into business enterprises, which they were permitted to operate in parallel with their university posts. In many cases, governmental research institutes were 'privatised', as prescribed by the free market gurus.

Nevertheless, laboratory life went on, almost unchanged. To be sure, the '80s and '90s were notable for the development of revolutionary new technologies for communication and computation. These transformed the technical facilities

[22]   ABRC (1987). *A Strategy for the Science Base*. London, HMSO.

and theoretical methodologies at the coal face of research. More time was now spent tapping keyboards and watching visual display units, than in turning knobs, reading meters or observing real phenomena. The internet enables long-distance data transfer, information retrieval and real-time collaboration by researchers at several centres. Research by *ad hoc* networked teams became a normal mode of knowledge production, in academic science as well as in technoscience.[23]

It was not just technoscience that was being 'globalised'. In fields such as ecology and astrophysics, academic research programmes were not only networked without borders: they were also organised transnationally. National research systems handed over some of their funding responsibilities to international bodies such as CERN (for elementary particle physics), EMBO (for molecular biology), and HUGO (for the human genome). As in the times of every past generation, science had transformed itself, materially and intellectually, by its own progress.

Nevertheless the everyday conventions of academic culture were still very largely maintained. Outwardly its social structure was unchanged. People still pursued highly specialised careers and worked in specialised research groups. They still did their best to follow the traditions regulating individual employment and preferment, the unwritten rules governing relations between juniors and seniors, the established customs ruling teaching and learning, the courtesies due to visiting colleagues and the attitudes appropriate to obnoxious bureaucrats. Young researchers continued to delve enthusiastically for knowledge whilst their seniors conceived exciting new projects and vied with one another for the means to undertake them. And yet, on larger scales of effort, in the connections between its more distant parts, over longer periods of memory and prospect, the knowledge-producing machine

[23] Gibbons, M., C. Limoges, *et al.* (1994). *The New Production of Knowledge*. London, Sage.

was being subjected to stresses that could only be accommodated by distortions, fractures and bodged repairs.

From the viewpoint of society these internal stresses and strains were invisible. All that could be seen of science was that it was making remarkable progress. It was not just that wholly new aspects of the natural world were being revealed. Research was also clarifying the underlying mechanisms of very old fields of knowledge, especially in biomedicine. The mapping of the human genome, for example, was the culmination of an immense scientific effort that had been going on for half a century. So it would soon be quite feasible to work out in principle how to achieve any desired practical effect — the cure for a disease, a novel organism, a semi-conscious mechanical device. The tortuous road of blind experimentation, speculative invention, ponderous research and costly technological development would be bypassed. Guided by academic theory, technoscience was about to become very like engineering.

This was what the scientists had been promising for all these years. Quite understandably, the public, their elected representatives and their governments were beginning to get impatient. They could wait no longer. They had tried flogging the old horse harder; and now the temptation to look for the golden eggs *inside* the goose was becoming irresistible.[24]

Post-academic Science

The barriers between the various modes of knowledge production are disintegrating. Academic science and technoscience are converging and interpenetrating. Does this mean that they are merging? They always drew from a common store of knowledge, techniques and personnel. Now, it looks as if they are being amalgamated into a single system.

To be more precise, it looks as if academic science is being taken over and reconfigured as an adjunct of technoscience.

[24]   Ziman, J. M. (1994). *Prometheus Bound: Science in a Dynamic Steady State.* Cambridge, Cambridge UP.

The evidence for this lies all around us. Book titles tell us that higher education is being commercialised,[25] and that science (meaning academic science) has become a commodity.[26] Governments are advising universities to replace state patronage with industrial research contracts. Praise and hidden subsidies are lavished on start-up firms created on the side by university professors. Agricultural institutes are 'privatised' and then immediately bought up by multinational companies. In the biomedical sciences, where there has always been a delicate balance of power between the interests of academic scientists, practitioners, and patients, so much of the research effort is now in the hands of profit-seeking enterprises that it took the combined authority of two governments and a very large foundation to even begin to wrest the map of the human genome from their grip.[27] There are many other examples.

In some circles, indeed, this development is welcomed with enthusiasm. We are told that the prosperity of the nation — of the great globe itself — will be fostered in future by a 'Triple Helix' of academia, industry and government. An ever-closer union of their research activities will open the way to a cornucopia of technological goodies — for example, the wondrous products anticipated by the state-sponsored 'Foresight' exercises in which their representatives already take part.[28] In effect, commercial techno-science is the only game in town. It should henceforth be enthroned as the only legitimate occupant of the place of science in society.

Do I exaggerate? Not really, if we are to believe some of the more enthusiastic effusions of its proponents. I am even astonished at their failure to submit their model to serious

[25] Bok, D. (2003). *Universities in the Marketplace: The Commercialization of Higher Education*. Princeton NJ, Princeton University Press.
[26] Gibbons, M. and B. Wittrock, Eds. (1985). *Science as a Commodity: Threats to the open community of scholars*. London, Longman.
[27] Olson, M. V. (2002). 'The Human Genome Project: A Player's Perspective.' *Journal of Molecular Biology* 319(4): 931-42.
[28] Irvine, J. and B. R. Martin (1984). *Foresight in Science*. London, Francis Pinter.

sociological, economic or political analysis.[29] They say that the purpose of the Triple Helix slogan is simply to promote more active cooperation between the major segments of the research system. But they entirely and seriously ignore the societal context in which this is to take place. They take for granted that the social role of science is entirely instrumental and ally themselves with the technocratic elements of free-enterprise capitalism. So the Triple Helix supports the dominant political agenda of our society. It also chimes perfectly with neo-classical economics. The cooperation it promotes is to be realised through the creation of joint ventures, but that is just small firms producing knowledge for sale. If these eventually fall into the hands of corporate industry, will that be an occasion for worry on an even larger scale? They will argue that it is no more than just the way the market produces the best of all possible worlds.

But even as an organisational model for technoscience, the Triple Helix is naïve. A very long succession of processes is required for a serendipitous discovery to become a marketable innovation. One of the weaknesses of the 'linear model' was the assumption that the linkages between these processes operate only in one direction from small individual discovery to large-scale commercialisation. Nevertheless, that doesn't mean that they ought all to be bundled together. 'Blue skies' research in academia is a very different social practice from the applied science performed in an industrial laboratory, which is different again from the testing of prototype products by a government regulatory agency. As we have seen, post-industrial firms have learnt to keep these functions apart, and even to 'outsource' them into separate enterprises. Certainly, it is essential to keep knowledge flowing back and forth along the chain. But it is questionable whether the encouragement of 'hybrid' organisations is the most effective way of bridging these gaps of style and function.

[29] Salomon, J.-J. (1999). *Survivre à la science: Une certaine idée du futur.* Paris, Albin Michel, Shinn, T. (2002). 'The Triple Helix and New Production of Knowledge: Prepackaged Thinking on Science and Technology.' *Social Studies of Science* 32(4): 599-614.

What is missing from the Triple Helix model of science is any consideration of its non-instrumental roles. Even by its own criteria, the model won't work. It takes the contribution of university-based research to be, at most, pre-instrumental, preferably guided along utilitarian lines by eerily prescient foresight. There seems to be no realisation that technoscience itself depends in a number of ways on its co-existence with a genuinely non-instrumental partner.

As we have seen, this 'enabling' role was traditionally filled by academic science. The Triple Helix discourse apparently assumes that this mode of knowledge production still exists. Thus, it takes for granted that scientists and their managers can tap into a self-correcting framework of reliable theoretical knowledge, a source of trained researchers, and a seemingly inexhaustible well of unexpected discoveries just waiting to be exploited. So it fails to ask whether academic science might have been so radically changed by its incorporation in the helix that it can no longer carry out its duty of care by its conjoint twin.

Some very experienced science watchers assert that the traditional academic mode of knowledge production — 'Mode 1' — is giving way to quite a new activity — 'Mode 2'.[30] But much of their evidence for this is beside the point. As we have seen, academic scientists nowadays are often 'networked into heterogeneous, interdisciplinary teams funded from diverse sources to operate within shifting paradigms and specialty structures'(to put it briefly). But these supposed characteristics of 'Mode 2' are merely signs of normal methodological progress. They say nothing about the purpose of the research, nor of its wider social context. This is just the way that all 'good science' is now done, in academia, government and industry, in the public, private, or not-for-profit sector, whether to gain 'CUDOS' from ones peers, to satisfy the requirements of ones post, or to make a profit for ones firm.

----

[30] Gibbons, M., C. Limoges, *et al.* (1994). *The New Production of Knowledge*. London, Sage.

On the other hand, it is simply not the case that all scientists these days are engaged (to telescope into a phrase several pages of sociological discourse) in 'problem-solving in contexts of application, guided by practical, societal, policy-related concerns in a range of environments already structured by application or use'. Such may well be said of professional research workers who are directly employed to produce 'useful' knowledge by commercial firms and government agencies. In other words, the other features ascribed to 'Mode 2' make it little more than a formal description of technoscience in its contemporary post-industrial form.

But this is not the present situation for all scientists. It does not apply to the hundreds of thousands of them, all over the world, who work in universities, in many state-funded research organisations, and in not-for profit foundations. It is true, as we have seen, that there are strong trends in that direction. Considered as 'modes of knowledge production', the sciences and their associated technologies have certainly changed a great deal in the past twenty-five years. The authors of this analysis state at various points that 'Mode 1' (academic science) still coexists with 'Mode 2' (technoscience). But they also believe that it is effectively outmoded, and will eventually be 'incorporated into this larger, more dynamic new system'.[31] To my mind, this is an oversimplification. It would be more accurate to say that research scientists in academia and elsewhere have abandoned many of their primeval 'Mode 1' practices, but that they have not all relaunched themselves in postmodern 'Mode 2' scenarios. Academic science is not being violently revolutionised, but neither is it being silently merged with, engulfed by, or simply superseded by the latest version of instrumental technoscience. But it certainly is passing through a period of radical cultural evolution, which should be watched carefully.

A distinctive 'mode of knowledge production' has developed or is developing in academia. But to designate it as

---

[31] *Ibid.*

'Mode 3' science would beg several theoretical questions. I prefer, therefore, to call it *post-academic* science, indicating both the break from and the continuity with the academic tradition.[32] This term also suggests that it has shed the intellectual aloofness that used to try to keep it at arms length from technoscience. Indeed, although it hints at 'post-modern' scepticism, this terminology suggests that it is actually a social form of endeavour with genuine cultural roots in the 'post-industrial' society where we now (so it is said) all live.

So what is the actual place of this new form of science in the contemporary social order? How do people relate to it, and how does it relate to people? Is post-academic science a stable institutional form? How well does it fulfil our ever-changing social needs? Does it, for example, satisfy the non-instrumental functions that lie outside the scope of 'Mode 2' technoscience? These are just some of the questions to which we shall now turn.

[32]   Ziman, J. M. (1996). 'Is science losing its objectivity?' *Nature* 382: 751-4, Ziman, J. M. (1996). 'Postacademic science: Constructing knowledge with networks and norms.' *Science Studies* 9: 67-80.

EIGHT

Post-Academic Science as a Social Institution

The science of today

Let us now, at last, look contemporary science squarely in the eye. What do we see? Much the largest part of it — commercial and state-run technoscience — is just straight-forwardly proprietary and utilitarian. To find out about it, you only need to pick up the technical language by training as an engineer, say, and taking a few courses in business management and market economics. That should be enough to understand the talk that is going on inside the corporate laboratory and trace the causes and consequences up until you get to the office of the vice-president for research and into the company board room. This is a cultural domain where hard cash and practical rationality rule — or are deemed to rule — without much respect for the past.

The less instrumental forms of knowledge production are much more difficult to observe clearly because they are undertaken by a great variety of social institutions. Some of these are venerable universities with centuries of dedication to teaching, learning, education, scholarship and research. Others are ultra-modern state-supported institutes probing the depths of nature with ultra-modern instruments. Others again are research laboratories funded by charitable foundations to study fearsome diseases. The most that they have in common is that they are all 'not for profit' organisations broadly associated with 'Academia'.

In the official statistics, their scientific work is classified as 'academic'. But as we have seen, it has changed so much in recent years that we should now call it *'post-academic'*. Nevertheless, its procedures are still hidden beneath the trappings of tradition. Established practices have retained their outward forms long after they have been modified to perform quite new functions. So our understanding of its place in modern society is obscured by the ideological smoke screen that still hangs around it.

The simplest way of explaining how it now operates is to consider in turn each element in Robert Merton's account of the normative structure of science.[1] Although post-academic science no longer conforms systematically to these norms, they conveniently span the various social practices required to produce (more or less) reliable knowledge. These headings also cover the features necessary for science to perform its more general non-instrumental roles in society. Thus, we can begin to see in detail just how well post-academic science fits into the global, pluralistic social order of our day.

Communication

As we have seen, formal communication is one of the defining features of science. In academic science, this was strictly limited to what was accepted by and could be cited from the archival literature. Post-academic science retains and draws heavily upon this procedure. Indeed, its customary practices are even more rigorously enforced than ever. A research claim must be submitted in a highly conventional style, in a stereotyped format, and subject to searching peer review before it is published. Public disputes over authorship, reviewer bias and priority are frequent. Charges of plagiarism, abuse of confidentiality, fabricated data and

---

[1]   Merton, R. K. (1942 [1973]). The Normative Structure of Science. *The Sociology of Science: Theoretical and Empirical Investigations*. N. W. Storer. Chicago IL, U of Chicago Press: 267-78.

other forms of fraud often disrupt the peace of academia. No doubt there are legal firms that specialise in such legalities.

And yet, the archival literature is being bypassed by other means of communication. Electronic networking not only speeds up the transmission of 'private' information within research teams. It also means that new findings become *publicly* available, over the internet or as widely distributed 'preprints', before they have been assessed by reviewers and sometimes almost before they have been verified by their authors. Strictly speaking, these are 'informal' communications, and therefore have no standing as 'scientific knowledge'. Indeed, they frequently turn out to have been quite mistaken, and are quietly withdrawn.

For scientists in the field, the fact that this type of information might well be unsound is not a serious problem. Even archival knowledge can often be unreliable. It is essential to have the latest news of where the action is. Along with all the other considerations that go into the making of research plans, they are assessed for their credibility and acted on accordingly. The post-academic research cycle is notably speeded up by this open 'networking' of preliminary results.

But this also makes public much that academic science customarily kept to itself. Indeed, despite all the concerns about the 'public understanding of science', the last few years have witnessed a great flowering of the 'popularisation' of science. Well-informed, brilliantly written books, articles and media programmes present all the latest discoveries in reasonably intelligible form. The trouble is (as scientists bitterly complain) this material is not policed by the official gatekeepers of scientific knowledge. Tentative research findings on contentious social issues are very quickly picked up by activists and trumpeted loudly in the media. The possibilities for commercial profit are scented, and blown up into speculative bubbles. Journalistic commentators — and scientific pundits, too — enlarge upon the moral, economic, social, medical, etc. implications of the latest 'discoveries' just as if they were now sober realities.

In vain do the scientific authorities plead for a pause until the official machinery of peer review has done its job. Was not this new and valuable knowledge produced by professional scientists, using scientific methods, scientific techniques, and scientific arguments? Surely it must be true — or at least much too true to be brushed aside. Where now are all the weasel words about possible 'experimental inadequacies', 'statistical uncertainties' and 'theoretical misconceptions'? Imputations that, perhaps, those stuffed-shirt referees have a finger in the pie, or are displaying a class bias, or are under orders from the government, or don't want their past errors exposed, or are just out of touch with the New Age. So the debate descends to something near to name-calling and conspiracy theories, and this is even before the central facts of the case are clarified.

It is equally unwise, however, to cling to the popular belief that publication in a peer-reviewed journal sets a seal of authenticity on a research claim. Academic science has always taken this as just a whistle stop on the route towards the goal of absolute truth. Scientists are pressed by anxious politicians and business-men for some way to measure of the credibility of a scientific finding. They have seized upon this favourable sign of communal acceptability as a reasonable bench mark of validity, without explaining that it is never a final judgement.

Indeed, much of the coffee-table talk amongst post-academic scientists is about the deficiencies of peer review. Frustrated in their ambitions to get their work published in high-impact journals, the scientists complain about the ineptitude with which their work has evidently been reviewed. This comes from the very people who perform this communal duty — but wearing different hats of course. Nevertheless, systematic investigations have largely confirmed the anecdotal evidence that peer review is a very uncertain process. Even the most prestigious journals and most distinguished academic publishing houses occasionally publish papers and books that are deeply flawed scientifically. Furthermore, the 'peer-reviewed literature' is not a closed set with clearly-defined boundaries. It extends out and down into innumerable esoteric, and sectarian corners of the scientific world, with diverse criteria of factuality and proof. So it is not difficult to find in some places in almost every paper some exaggerated inferences from unreplicated observations or experiments, which can be called forth to support almost any half-baked theory.

The body of knowledge that post-academic science actually opens to society is thus of very uneven quality. Some of it is of almost impeccable credibility: some is just tosh. One can no longer assume, as the academic ethos was supposed to guarantee, that it was all reasonably reliable, or at least that the researchers themselves would explain why it was still somewhat uncertain. Historically speaking, this faith in the 'validity' of scientific knowledge may never have been justified. But, as we have seen, that was the myth that enabled academic science to carry out many of its non-instrumental social roles.

Doubts about the credibility of post-academic science thus weaken its performance of these roles. These are not just 'philosophical' doubts. They are not, as is often asserted, induced by reckless post-modern criticism.[2] They reflect genuine uncertainties, contradictions and ambiguities in the information available to the public under post-academic conditions.

With experience, of course, one can soon learn how far to trust any particular item in this information. A lot still depends on the scientific reputation of its source. Communications from several independent sources deserve serious attention. But when such prudential tactics fail, it is tempting to look behind the scenes.

Academic science throws a veil of ignorance over all the actual work that goes into the writing of a research paper. It is sometimes asserted that the 'scientific attitude' obliges researchers to tell absolutely everything they know or suspect about their field of study. This is not the case. The norm of 'communalism' only requires them to produce all the evidence they consider relevant to the particular discovery that they are publicly claiming. All that matters is that sufficient

[2]   Gross, P. and N. Levitt (1994). *Higher Superstition: The Academic Left and its Quarrels with Science.* Baltimore MD, Johns Hopkins UP.

information is given for it to be convincingly shared and replicated by other scientists. They are not bound to make public all the false starts, instrumental breakdowns, faulty data, and inconclusive experiments that make their days so long and laborious. It is part of their professional expertise to recognise and exclude what is not, after all, a serious contribution to knowledge. Only the historian of science, half a century later, is permitted to piece together the course of events out of the tattered laboratory notebooks, the occasional private letters, or the imperfect reminiscences of the survivors and their fragments of the past.

By contrast, post-academic administrative procedures, backed sometimes by legal authority, often require the research process to be recorded systematically, in case it needs to be accounted for, day by day, data point by data point, as if in the ledger of a bank. This is not a fanciful analogy. Not infrequently, a non-instrumental investigation unexpectedly reveals the way to a valuable practical discovery. In that moment, an intangible item of knowledge has become a piece of intellectual property, a commodity with a price tag. It must be immediately seized upon, appropriated legally, and then exploited for all it is worth. The event must be recorded carefully, for it may well be the key point in the award of a patent. But just like an entry in a bank account, it must also be kept secret, lest it be put to use by a rival. So the post-academic researcher has to allow for the possibility both of enforced disclosure and of enforced 'enclosure' of his or her professional work.

The plain fact is that post-academic science is not insulated institutionally from its technoscience twin. It has become quite common, for example, for universities to carry out contract research for commercial firms or government departments. The results of this research are, of course, simply reported in confidence to their lawful owners, just as if they had been obtained in the company or departmental laboratory. If they become public at all, it is in a form that has been carefully sanitised. Indeed, it has been known for a government department to initiate and support a research project and then to shoot itself in the foot by clumsily trying to suppress or launder findings that contradict its avowed policies.

More insidious, perhaps, is the post-academic practice of encouraging a business firm to 'sponsor' a research institute in return for privileged access to the knowledge it produces. The arrangements for this are very varied, and are often strongly criticised by alert faculty members. Perhaps they are innocent enough, and cause no significant delays in getting the work into the public domain. But even if they can be presented as fair protection for sensitive commercial information, or a very modest recompense for a very generous act of patronage, they always rouse a suspicion that the whole truth is not being disclosed. When this shadow falls across a whole university it is clearly in big trouble.

Post-academic science faces researchers with uncomfortable personal decisions about the communication of their discoveries. Should an exciting new finding be published in a scientific journal so it can figure proudly in ones CV? Or should it be kept quiet enough to be exploited financially as intellectual property? Is the ultimate aim a Nobel Prize; or is it to make a billion dollars as a technoscientific entrepreneur? The imaginative literature about the vicissitudes of the scientific life is only just beginning to catch up with such dilemmas.

In practice, of course, as usual, people and organisations learn how to paper over such contradictions and live with their disturbing effects. And the data and conjectures that are thus temporarily kept out of view are seldom of any lasting scientific significance. But this disjunction in the most fundamental social practice of post-academic science — the public communication of research claims — is an outward sign of an internal moral rift that puts all the knowledge that it produces into doubt, and seriously affects all its social functions.

Orientation

Academic science has always sought to be *universal*. Its foundational thinking was on the common ground of human existence; but it looked out on the cosmos and built for eternity. Post-academic science is more modest and down to earth. Perhaps its lofty aspirations really have been diminished by the post-modern critique. Or do the physicists simply not care that their notion of a grand 'Theory of Everything' is completely discredited elsewhere in the postmodern intellectual world. More likely, researchers in every field have been humbled by the way that the advance of knowledge has revealed alps upon alps of complication rather than the prairies of law-like simplification that they were hoping to find. It is not that all complexity is chaotic, and therefore beyond explanation. It is that even that such an apparently intelligible system as the human genome project is far more variegated in detail than we had suspected, because it is far less amenable to generalisations than we had hoped. As we shall see, most scientists now realise that a *pluralist* orientation is essential if they are to make any sense of the knowledge they produce in such abundance.[3]

The social harness of post-academic science pulls it in the same direction. Researchers are continually being reminded that their work is actually supported by the public very largely for its pre-instrumental promise. I am not referring simply to the percolation of contract research into the innumerable small worlds of academia. The whole rhetorical atmosphere surrounding science is saturated with utilitarianism. The prospect of profitable application, however distant and conjectural, is what supposedly gives research its true value. Many projects are undertaken because they are 'good science', and are not seriously expected to produce any exploitable results. Nevertheless, they have to be formulated as if they were somehow, in a vague 'strategic' sense, steps towards that goal.

Indeed, whole disciplines are relabelled in that spirit. Biologists, for example, are ready to call their research 'bio*medical*', or 'bio*technological*' even though they are not

---

[3]   Ziman, J. and M. Midgley (2001). 'Pluralism in science: a statement.' *Interdisciplinary Science Reviews* 26(3): 153.

really interested in, or even able to articulate, any practical application for what they hope to discover. Some of this is just opportunism. But the overall pattern of state patronage for post-academic science shows that hard-headed legislators and government officials strongly favour apparently utilitarian categories. For example the relative distribution of funds within basic biomedical research, broadly reflect the *classes of disease* of greatest public concern.

Even such big spenders as high-energy particle physics and astronomy are no exceptions to this. Despite their manifest uselessness, these more sublime sciences are largely funded because they are somehow associated in the public mind with the gruesome but supposedly necessary capabilities of nuclear and space weaponry. Post-academic researchers in these fields are careful not to mention this connection in public. On the one hand, if the post-academic argument were true it would call down on them considerable moral indignation. On the other hand, if they were to argue strongly that it is all untrue then they might well call the bluff on their own demand for funds.

The spirit of post-academic science thus favours the detailed study of restricted areas of knowledge. This naturally involves the solution of relatively 'local' problems of fact, theory and technique. These problems may be quite abstract and theoretically challenging. They are frequently far removed from any plausible 'context of application'. Nevertheless, their solution can be of great value to technoscience by making it possible to plan research and invention in order to achieve preconceived practical ends. In the jargon of science studies, a well-established *paradigm* — that is, a reasonably reliable 'map' of an area of knowledge — permits the *finalisation* of research programmes — that is, the realistic planning of projects to achieve foreseen theoretical and practical goals.[4] In the past post-academic science gave top priority to the investigation of the genetic role of

[4]   Jagtenberg, T. (1983). *The Social Construction of Science*. Dordrecht, Reidel.

DNA, because this opened the way to the systematic mapping of various genomes and to the possibility of actually 'engineering' living organisms. The placing of 'genomics' in larger schemes of knowledge of the living world is clearly of much less urgency.

This tendency is also favoured by technical and methodological factors. Post-academic science has not been slow to appreciate the efficacy of the 'Mode 2' style of research. Many problems can only be tackled by *heterogeneous* networking — that is by temporary teams of researchers recruited from several different disciplines working on several different research sites in several different institutional settings, some of which may not even be inside the 'academic tent'. To formulate such a project and to assemble and coordinate such a heterogeneous working group requires a specific objective and a plan of action that seems likely to produce reasonably significant results. In other words, it is only likely to get under way if it is directed towards the solution of a well-defined problem in an area of knowledge and technique that is not still shrouded in mystery.

To put this the other way round. Post-academic science does not much encourage relatively small-scale speculative ventures, even by well-established scientists, into highly uncertain and enigmatic areas of research. This was always one of the strong points of academic science. Needless to say, it did sometimes enable the production of a great amount of bunkum, and its few shining achievements were too often only recognised posthumously. But as we have seen, one of the non-instrumental social functions of science is to keep on thinking the unthinkable, just in case it becomes real. Is post-academic science just a bit too pragmatic and socially expedient to sustain this role?

Of course, some fields of big science seem still to preserve the magical tradition of academic science. One can only admire the nerve with which the astrophysicists lobby for vast sums to be spent on exquisitely sensitive instruments for (possibly!) observing the universe by the light of (never yet detected) gravity waves, whilst their rivals, the particle physicists undertake a billion-dollar experiment to pick up faint traces of hypothetical particles that may never have existed in the universe before, and which might—just might, we'll never know if it does—destroy planet earth itself in a flash.[5] But these are, so to speak, licensed domains of scientific entertainment, and quite disconnected from anything human.

Administration

Academic science did not know the meaning of 'management'. All that aspect of life came under the heading of 'administration', and was derided and minimised by 'real scientists'. For them, academia was a playing field, a market place, even a battleground, whose boundaries, goal posts, resources and financial frameworks were maintained by others. Apart from serving an occasional term as Dean of a faculty or Pro-Rector, they were free to do 'their own work' as they said, teach their courses and wrangle on equal terms with their colleagues about parking privileges and examination curricula. Even the legendary powers of a professorial head of department were limited to hiring and promoting junior colleagues most of whom he could not later either fire or even demote.

Post-academic science is much more highly organised but not significantly more hierarchical. Its communication networks, instrumental facilities, inter-institutional team arrangements and service infrastructures require elaborate social structures in which the researchers themselves are professionally involved. But they still refer to these managerial responsibilities and controls as 'administration', and try to keep them from taking up too much of their time and so damping the intellectual ardour of both themselves and their subordinates.

What has really changed, though, is the role of *money*. When they are not swapping yarns about the iniquities of peer review, post-academic scientists are talking about

[5]   Calogero, F. (2000). 'Might a laboratory experiment destroy planet earth?' *Interdisciplinary Science Reviews* 25(3): 191-202.

'funding'. This is now the standard term for the rows and rows of figures in their project budgets, departmental accounts and the final grant proposals.

Of course that's right. The system could not work without rich flows of this life-giving fluid through its arteries. But how to maintain an adequate supply of funds absorbs a lot of the time that could be more profitably devoted to research and frequently becomes as worrying to researchers as how to solve their scientific problems. Furthermore, this anxiety chains them to the sources of these supplies quite as effectively as to any formal authority. They are rather like most peasant proprietors — nominally free to do what they like, but too heavily in debt to landlords and money-lenders to be genuinely self-sufficient or independent in what they produce.

So post-academic science is always in the hands of its external paymasters. The researcher is continually in thrall to the funder, just as the farmer is to the bank. And just as the bank can refuse, for its own reasons, to lend the capital requested for a proposed development, so a funding body does not have to justify publicly its decision not to award a grant for proposed research project. This power to refuse support, exercised in private, is not less potent than open control.

In practice, of course, a researcher with a good track record as a 'principal investigator' can keep up enough financial momentum to stay in the game. A typical survival strategy is to hedge ones bets. By acquiring a number of assistants, one can form a research group that entwines a number of projects into a continuous braid. The reins of control are also held very loosely. As we have seen, the official policy of many funding bodies is simply to support 'timeliness and promise', as defined by expert reviewers and panel members. So a project that accords reasonably well with current paradigms of 'good science' is not likely to be impeded.

Even then, expert opinions can differ widely. But so can the putative objectives of a given research plan. So if a proposal fails with one funding source it may well succeed with one with a different mission on the other side of town. Indeed, one of the 'Mode 2' features of post-academic science is that the range of potential funders is broadened to include industrial and commercial enterprises as well as not-for-profit foundations. This is strongly favoured by governments, since it takes off their shoulders some of the increasing financial burden of maintaining the national science base. Academic institutions themselves often receive large state subsidies or have enough endowment income to cover at least a proportion of the research carried out by their employees. In some countries, on the other hand, the allocation of funds to specific research projects is still determined by fierce battles amongst the leaders of various groups inside the walls of a 'national research council' or 'academy', with relatively few links or leaks to the outside world.

All this is discussed interminably in the world of post-academic science policy. Scientific leaders, academic *responsables*, government officials, industrial managers, politicians and other 'decision makers' have a difficult job. How can such a ramshackle, loose-limbed system be made to adapt smoothly to the changing demands of the economy and the polity? Can it be protected from the excesses of bureaucratic zeal, financial accountability, economic efficiency, political expediency and doctrinaire populism? And what about all the detailed effects on personal careers, professional responsibilities, institutional policies, material facilities and work patterns?

But these are mostly technical issues, of interest only to those directly involved. The question for this book is: how does this way of funding post-academic science affect its place in society at large? The standard answer is that it is a very effective mechanism for producing large quantities of reliable pre-instrumental knowledge. On the one hand, the organisms that do research are fed generously and are encouraged to feel quite free to do their own thing, to follow the same scheme of thought. Meanwhile the financial leash keeps them from wandering far afield.

One could say that this is effectively the same as the 'Mode 2' practice of outsourcing exploratory work to specialised 'research boutiques'. Indeed, this mode of knowledge production permits technoscience to percolate freely into academia. When its research council grants run out, a post-academic research group can often keep itself going on commercial or governmental research contracts, almost without changing the substance or goals of its investigations. This, of course, is one of the ways in which the strands of the 'Triple Helix' are supposed to merge into one another.[6]

But what about the non-instrumental role of science in society? How is this affected by post-academic funding practices? Clearly, these make the researchers and their research much more open to 'external' influences. As I have argued at several points already, it is naive to suppose that any systematic activity can be entirely dissociated from the interests of whoever supports it materially. When this linkage is specific and direct, as when a particular research programme is approved in advance and financed by a particular funding body, can one really believe that it is not designed to further the interests of that body?

In the past, these connections were largely covert, and perhaps not very influential. Post-academic science now makes them much more explicit. It does society a service by revealing the various political and corporate interests that are operating within it. But the fact that these forces are becoming increasingly powerful is a serious cause for concern. There is now much anxiety in the scientific world about excessive commercial intrusion into academia. This is mostly about how pre-instrumental research should relate to full-blown technoscience, but it applies also to the more general role of science in society.

In sum, post-academic scientists can never plausibly claim to be effectively 'disinterested' in the idealised aca-

---

[6]   Shinn, T. (2002). 'The Triple Helix and New Production of Knowledge: Prepackaged Thinking on Science and Technology.' *Social Studies of Science* 32(4): 599-614.

demic sense. They cannot throw off the cash nexus that enables their work. The production of knowledge is an unusual 'industry' but it cannot escape entirely from its material base. As with every social formation, its economic aspects and implications should not be ignored or be cavalierly dismissed. Of course scientific and technological research has other cultural features. Even when presented metaphorically as an interlocking system of 'markets', it does not conform tamely to the narrow doctrines of neo-classical economics.[7] All the same, questions such as who pays for what — and why — have to be asked and answered. And the exposure of what old-fashioned Marxists call 'contradictions' can reveal conflicting forces that are not all working for sweetness and light.

One can see, for example, why generous state patronage of research is still strongly favoured by thoughtful students of science and technology policy.[8] This is not just because this subsidy is required to make up for market failure in the production of a public good — typically defined as the national knowledge base. It is also favoured because it attaches the researchers to the *public* interest. Government money laundered through more or less independent bodies run mainly by scientists can support research that is reasonably free from partisan influences, commercial and/or political. No doubt this is a transparent political device, easily deconstructed by critics on both wings. But experience has shown that it can be made to work well enough to protect at least a certain number of researchers from gross material interests that might conflict significantly with their non-instrumental social functions. As we shall see, this is a very important consideration in the preservation of these functions.

---

[7]   Ziman, J. (2002). The microeconomics of academic science. *Science Bought and Sold: Essays in the Economics of Science*. P. Mirowski and E.-M. Sent. Chicago, University of Chicago Press: 318-40, Ziman, J. (2003). 'The Economics of Scientific Knowledge: Review of Shi (2001).' *Interdisciplinary Science Reviews* 28(2): 150-2.

[8]   Nelson, R. R. (2003). 'The advance of technology and the scientific commons.' *Phil. Trans. R. Soc. Lond. A* 361: 1691-1708.

Gestation

One of the effects of the transition to post-academic science is to throw a great deal of emphasis on the *initiation* of the knowledge-production process. Despite the uncertainty of its outcomes, science is a highly *intentional* activity. Academic science called this activity 'research'. In principle, it is what 'researchers' do as they perform actions that they believe will generate new knowledge. These actions were not disconnected from the thoughts that occasioned or guided them. Together they made it evident that scientific 'creativity' was in operation.

Post-academic science makes much more of a mouthful of this process. It interposes a whole new stage between the thought and the action. The idea of performing a specific investigation has to be made explicit, not only to the researcher but to others, before it is actually carried out. The requirement that it had to be formulated as a *project* in order to be funded is thus a radical change in the practice of research.

This alters completely the metaphorical economics of the research system. As we have seen, academic science segmented its activities in terms of reports of work which were *actually accomplished*. These were 'marketed', so to speak, as more or less credible items of achieved knowledge. That is, they were presented as ready for use, even though they might quite possibly turn out to be quite unsound.

Post-academic science, by contrast, parcels out its activities in the form of proposals for work *still to be done*. These are marketed, typically for real money, as knowledge *futures*. Their potential for use, whether technologically or conceptually, is hypothetical. In strict accountancy terms, the outcome of a research project is as valuable as the proverbial 'pig in a poke'. It feels as if it might be worth the money, but we can't be at all sure until we've paid up, opened the bag and inspected our purchase.

This uncertainty is, of course, intrinsic to research. By definition, the 'production of knowledge' excludes the *repro*duction of what is already known. That is why science is more like prospecting for gold than the mass manufacture of costume jewellery. But as research becomes more and more heavily capitalised and collectivised, its financial backers seek more and more security for their 'investment'. So they push for more information about the 'prospects' of each investigation, as viewed from the moment when it was conceived. A clear account of its objectives and methodology ought to give a good indication of its likely outcome.

Indeed, many research projects are carefully planned, from the beginning, to produce results within a predetermined range. They may, for example, have been *commissioned* by a funding body to serve its own interests. As we have seen, *contract research* is typical of the regions where post-academic science overlaps and merges with technoscience. Such contracts are even put out to tender. The questions to be asked are prescribed by the funding: research groups who compete in terms of the technical expertise they propose to use, as they answer the questions.

If the appropriate questions are still somewhat obscure, funding bodies sometimes issue a *request for proposals*. That is, they designate a specific problem or problem area — usually, but not necessarily, 'arising in contexts of application' — on which they intend to focus a programme of research. Researchers then submit plans for particular research projects whose findings might lie in this domain. They thus compete not only in terms of their technical capabilities but also in terms of their grasp of the problem and how it might be tackled. They thus have an opportunity to demonstrate the personal *originality* prescribed by the academic ethos. Furthermore, as we shall see, this is often an excellent way of producing relatively non-partisan knowledge relating to a contentious societal problem.

But post-academic science tries to maintain its academic heritage by undertaking a substantial amount of research in the *responsive mode*. Grant-awarding bodies advertise deadlines for the reception of project proposals in very broad fields of knowledge. In due course, after an elaborate process of allocation to specialised panels, peer review, further expert scrutiny, etc. some applicants are awarded grants and can go ahead with their research. For the others it is a question of better luck next time.

Scientific activity is thus generated and shaped by a very curious social process. In effect, it is the outcome of a one-way interaction between the researcher who originally conceived the project and the group of reviewers who decided to fund it. In a way it preserves the strategic autonomy of the individual, if only within the bounded landscape of collective imagination. It leaves quite a lot of space for the display of personal originality, even though it limits the exercise of unconventional 'creativity'. So it conforms to the letter, but not quite to the whole spirit, of the academic ethos. Perhaps academic science, even in its heyday, was never much better than this at living up to its ideals.

Working scientists are all too familiar with practical defects of this process. It is the area where the slick judgements of ill-informed reviewers are most bitterly resented, not least because they directly affect the livelihood of the grant applicant. The amoral opportunism of the legendary Dr Grant Swinger is all too realistic. The system also facilitates genuine misdemeanours such as plagiarism and — in effect — cradle-robbing. Conflicts of interest are difficult to discover and control. Ever-expanding bureaucratic bumf impedes the real work of research. Talented scientists have to waste time writing proposals that are ultimately turned down.

The practical advantages of the system are not unappreciated, although seldom publicly celebrated. It is obvious, for example, that the method provides an orderly procedure for the detailed disbursement of substantial sums of public money. Difficult choices are made by well-informed persons, essentially on the basis of demonstrable merit. It keeps political favouritism at bay without giving excessive power to permanent officials. It thus fits neatly into our pluralistic polity.

It also upholds the quality of the knowledge it produces. The mind is wonderfully clarified by having to plan and justify a research project. Awareness that the proposal will be carefully scrutinised by expert authorities sets a high standard of substance and presentation. It obviously has to be written in the specialised language of its subject, with appropriate references to its scientific context. This is valuable training in thinking and expressing oneself in these terms. Techniques and methodologies have to be considered in advance, and not merely improvised and rationalised after the event. These are not trivial requirements. Researchers perform better if they are forced to make plain to others — and thus, reflexively, to themselves — what problems they are proposing to tackle, what questions they are going to ask and what the significance of the likely answers might be.

Nevertheless, many post-academic scientists are uneasy about handing over the keys of every gateway into the realm of new knowledge to groups of their 'peers'. It subordinates individual insights to collective opinion. The result is that genuine novelty is often suppressed. Unconventional interdisciplinary projects fall through the gaps between established specialties. On the other hand, fashionable ideas are amplified. As sociological critics frequently seem to argue,[9] knowledge of doubtful value is wilfully 'constructed' by copy-cat scientists who formulate projects that they know will be acceptable to the usual reviewers.

This is particularly serious in the human sciences, where it is often very difficult to separate a dominant scientific paradigm from a ruling cultural ideology. Some economists, for example, are very sceptical about the outright individualism on which almost all current work in their discipline is theoretically founded.[10] But they have difficulty in getting their research funded because it is interpreted as an 'irrational' attack on the basic principles of free-enterprise capitalism, which the great majority of their colleagues assume to be incontrovertible. The sociologists of science have a

---

[9]   Pickering, A. (1984). *Constructing Quarks: A Sociological History of Particle Physics*. Edinburgh, Edinburgh UP.

[10] Ziman, J. (2002). No Man is an Island. *Hermeneutic Philosophy of Science, Van Gogh's Eyes, and God: Essays in Honour of Patrick A. Heelan, S.J.* B. E. Babich. Dordrecht, Kluwer: 203-18.

strong case here for the influence of political doctrines on scientific thought.

In general, however, this is a very delicate issue. The 'collectivisation' of problem choice markedly raises the *average* standard of research. But in doing so it eliminates many marginal projects with very doubtful prospects of success. This is not foolish. Idiosyncratic scientific ideas are not necessarily superior to the conventional wisdom from which they deviate: on the contrary, they are frequently just as silly. And yet, as history continually proves, they are sometimes the knives that prise open the shells of complacency that encase most social groups. So perhaps they should somehow be permitted to survive.

The real trouble is not with peer review. It is that the actual formalisation of the research process into 'projects' is inconsistent with its innate uncertainties. In principle, nothing new can be started until there is a contract or a grant to fund it. Serendipitous opportunities cannot be seized because they were not foreseen in the project proposal. So all novelty, all innovation, emerges outside the frame designed somehow to produce it!

Post-academic science thus excels at what Thomas Kuhn called 'normal' science — the solution of scientific 'puzzles' and the investigation of recognised 'problems'.[11] But it does not truly welcome 'revolutionary' projects for resolving 'anomalies', probing 'enigmas' or overthrowing the *status quo*.[12] It is thus weakened in the performance of its traditional non-instrumental social function as a source of well-founded alternatives to the established opinions of its time and place.

---

[11]   Kuhn, T. S. (1962). *The Structure of Scientific Revolutions*. Chicago IL, U of Chicago Press, Ziman, J. M. (1981). *Puzzles, Problems and Enigmas: Occasional Pieces on the Human Aspects of Science*. Cambridge, Cambridge UP.

[12]   Ziman, J. M. (1981). *Puzzles, Problems and Enigmas: Occasional Pieces on the Human Aspects of Science*. Cambridge, Cambridge UP

Evaluation

Post-academic science is weighed down with *evaluation*. Researchers are expected to be *accountable* for their research and to be *assessed* for how well it is performed. Post-academic scientists not only have to get their doctorates, and compete for employment and promotion. They also have to submit their achievements to frequent, detailed public scrutiny by non-expert bodies. There always seems to be somebody looking over their shoulders. They never feel that they can relax and get on with the job — a job that they enjoy and do as well or better than most people could.

That is how it seems. But of course academic science was never entirely free from such stresses. The evaluation process was for real, even though its formal procedures were performed relatively discreetly, at intervals of many years. Quality control was concentrated in the hands of the senior 'peers' of each discipline or sub-discipline. But academic examiners and referees were expected to be well acquainted with, and responsive to, communal opinion.

In particular, academic excellence was supposed to be defined solely in terms of actual knowledge produced. All that really counted was research claims that had been approved for publication by reviewers, upheld successfully in public debate, widely referred to in the literature, and even so completely accepted that they were no longer specifically cited. These processes embodied a variety of tacit ideals of what constitutes 'good science', such as that it should be novel, technically expert, and to contribute significantly to the progress of the subject. But they were primarily *critical* in terms of its substance — i.e. that its results should appear 'reliable' by the standard criteria of the field.

Post-academic science is obliged to make these processes more transparent. Managerial doctrines insist that performance should be kept up to the mark by frequent monitoring by the relevant authorities. So expert assessments of people, projects and institutions have to be presented in quantitative terms (like the grade points awarded to students), so that they can be justified to non-expert outsiders. For example, publications and citations are counted year by year, weighted with journal 'impact factors' and aggregated with other equally arbitrary numerical measures into 'indicators' by which they are compared and contrasted, across time and diverse populations.

No doubt this policy has activated research in institutions which had given up participation in the knowledge market-place. It has also 'remotivated' — or taken out of the competition for scarce resources — tenured researchers who had been getting a bit stale and lazy. And the need to justify their work to outsiders has made scientists much more aware of the wider societal needs to which it might be relevant. Post-academic science is funded generously for the supposed benefits that will come from the torrent of incomprehensible knowledge that it produces. Both science and society must have well-founded confidence that this money is being employed to good purpose.

Nevertheless, some of the criteria by which research performance is now being assessed are only very indirectly related to its scientific quality. Societal merit is highly praiseworthy in principle, but how can we measure it in practice? In a money-crazed culture, cash value rules. So it is not surprising to find that individual researchers, research groups and institutions win high marks simply for their success in getting research grants. One can perfectly appreciate the very sordid reasons why a candidate for an academic appointment at a cash-strapped university should be expected to show that he or she can attract enough outside funding to support their research. But this could be contract work, of no intrinsic scientific merit, for an undiscriminating commercial firm.

In the end, however, these are relatively minor deficiencies. The real problem is that post-academic science stresses 'evaluation' at the expense of 'criticism'. Too much attention is given to the technical and managerial efficiency of the production process: not enough to the intellectual and moral quality of the product.

Let me set this in a larger frame. One of the vital social roles of science is to inject large doses of *scepticism* into our knowledge of the world and of ourselves. This is a norm that operates both through the questions to be researched and on the answers we think we have obtained. Post-academic science does not give itself over fully to this norm. It lacks the confidence to be ruthless in its demolition of unreliable research claims that happen to suit influential social interests. It is frequently weak-kneed in its opposition to dangerous nonsense and in the defence of awkward truths, especially when this would bring it into open conflict with the powers that be. What is more worrying, it is too dependent on those powers to encourage the questioning that might undermine them. It pursues technically excellent knowledge with undiminished vigour: yet it often seems to have lost sight of the values that made such knowledge worth pursuing.

Differentiation

Post-academic science is 'cosmopolitan', rather than 'universal'. It welcomes participation by scientists from a diversity of research traditions and is good at getting them to work together on common problems. Research by *multidisciplinary teams* in the 'Mode 2' spirit is typical, even on problems arising in disciplinary contexts far from potential application. Thus, for example, fundamental research on the relationship between thoughts in the mind and neurophysiological activity in the brain requires the combined efforts of specialists in a whole range of disciplines, from cognitive psychology to electronic engineering.

Nevertheless, the differentiation of individuals into innumerable specialties is still the norm. The major disciplines of academic science are still the homelands where researchers are trained, where they teach, and where they claim professional allegiance. Even though I have not written or spoken a word of new physics research for the past twenty years, I am still considered to be a *physicist* — and even a bit by myself.

It is true that the frontiers between the traditional sciences are now broken up into numerous semi-independent interdisciplinary border states. Powerful research techniques trail like continental superhighways across the whole intellectual map, so that the study of mediaeval manuscripts benefits from computational methods invented to decode the human genome, and dinosaur bones are dated by their radio-activity using the knowledge of new isotopes. But the map is not a continuum. It is still a mosaic of localities, defined by the highly specialised bodies of knowledge and/or technical skills commanded by its inhabitants. If I were still in physics, I would be even more narrowly identified as a 'disordered condensed matter electron transport theorist' — than I once was.

Furthermore, the differentiation of post-academic science into 'interdisciplinary sub-disciplines' is not just a continually shifting scene. Certainly paradigms, problem areas, sub-fields etc. combine, bifurcate, recombine, reconfigure at a bewildering rate. But they seldom actually die the death. More often, what starts as an *ad hoc* multidisciplinary 'workshop' on an emerging scientific problem crystallises into a semi-permanent 'institute' or 'research centre'. This may lie, or be organisationally networked, across the boundaries between academia, government and industry. And to get funding some at least of its research projects have to be formulated instrumentally. That requires the inclusion of researchers principally engaged in technoscience — which is also, of course, equally finely divided into technical specialties. So post-academic science is typically heterogeneous in the composition of its research teams.

One might think that these are just managerial issues, irrelevant to the function of science in society. But as we have seen, this often requires a wide-angled perspective on the world. Post-academic practices do *not* encourage breadth of mind. The expertise of individual post-academic scientists is often extremely narrow. Team work enables them to contribute collectively to the production of knowledge of very wide scope — for example, global climate change. But they can each play their part in this without taking off their specialist knowledge blinkers. It is true that the scientists quite frequently write about their work outside the mainstream journals of their disciplines. But this usually takes the form of a series of a disconnected chapter in a multidisciplinary 'symposium' volume compiled incoherently out of the papers presented at a conference or 'workshop'.

There is simply no incentive to produce *transdisciplinary* knowledge. The best post-academic scientists are so preoccupied with getting research grants that they seldom engage enthusiastically in undergraduate teaching. Radical curriculum reform requires years of thankless academic politicking, so their lecture notes and textbooks just display the latest research findings in traditional disciplinary frames. And there is not much project-funding for spending ten years ahead, or for focussing on an unacknowledged enigma, or just reflecting upon a whole field of knowledge, fitting together the fragments of what is supposedly known until (possibly) they reassemble in new shapes, and reveal unsuspected patterns in the state of things entire.

It could be said that this reflects the influence of post-modern criticism, which rightly discourages the production of overarching theories and 'grand narratives'. Post-academic science has become suspicious of facile reductionism. Certainly, the more we learn about human beings, their social institutions, the biosphere, the history of the Earth, micro-physics, the nature of the cosmos itself, the less likely it would seem that the whole caboodle can be neatly wrapped up in a single theoretical parcel.[13] But there is also a more mundane explanation. Scientists nowadays cannot afford to be 'generalists' and have even less opportunity than ever to develop the prize of *wisdom*.

Occupation

Post-academic science still offers all the old personal gratifications and also the discontents to those who begin by fancying it and discover, with great pleasure, that they are reasonably good at it. Young people still become entranced

[13]   Ziman, J. (2003). 'Emerging out of nature into history: the plurality of the sciences.' *Phil. Trans. R. Soc. Lond. A* 361: 1617-33.

with science, devote years to the study of a particular branch of it, chain themselves to a doctoral topic, suffer low pay for long hours of seemingly fruitless drudgery, luxuriate in the exercise of technical virtuosity, delight in occasional discoveries, enjoy the companionship and esteem of their colleagues, negotiate cagily with competitors, bite back chagrin at being forestalled, become bitter at being passed over, develop inordinate vanity in success — and all the other vicissitudes of an outwardly straightforward life.

What is surprising is that this cheerful vocational spirit is still very much alive in a relatively harsh professional environment. It is true that career paths are quite diversified. Job opportunities for trained post-academic scientists are not limited to research universities. If one is unhappy in there, one may find employment doing much the same type of work in a non-university institute, research centre, government agency, industrial laboratory, or a newly developed industrial estate of a multinational company. There are so many choices from a networked project, small hi-tech firm, charitable foundation or government institution. None of these pay their researchers very highly, but then there is always the possibility of striking out on ones own and making a few million dollars as a technoscientific entrepreneur!

Unfortunately, this apparent diversity of career openings only accentuates the insecurity that mars post-academic science as an occupation. Yes, of course, research is a very uncertain activity. It is a domain where chance continually operates, even if it especially favours the prepared mind as Pascal told us two centuries ago. Even 'normal' science, which seems to be so very level, has its ups and downs. There is indeed a sense in which any professional researchers who could not take risks would be useless. They need to rely on sufficient stability of employment which could carry them patiently and scrupulously through quite long periods of sheer bad luck when the hoped-for breakthrough does not occur.

And yet post-academic science systematically flouts this elementary principle. Insecurity is built in it from the beginning. Formal professional status — e.g. a doctoral degree —

takes long to acquire, and is not attained at all by quite a number of those who set out to achieve it. This is usually followed by a succession of post-doctoral positions on short-term contracts, a state of uncertainty and intellectual servitude that is prolonged for many until they are old enough to retire. Even those who are rather successful scientifically may not have legally irrevocably stable employment. And in institutions where the blessed state of permanent 'tenure' still exists, it may come only when its recipients are approaching middle age, often when they are already past the peak of evident excellence that is attested publicly to their much above average competence.

The 'treat 'em mean and keep 'em keen' school of management wields the stick of unemployment. But most scientists are already highly motivated by the carrot of communal esteem. Their competitive instincts are not further roused by anxiety about where they are going to get the money that they and their colleagues need for the next round of research. This not only takes up a great deal of time and effort. It also forces senior researchers and research groups to plan their scientific work opportunistically rather than strategically. A career path that ought to be continually focussed on the production of knowledge is diverted towards the filling in of forms and the consumption of funds.

Post-academic science is good at fostering cooperation and organising team work. It exploits the multiplicity of our abilities, our willingness to learn new skills and our innate capacity to connect with our fellows.[14] It can be a most satisfying occupation for clever, reasonably sociable people who enjoy 'problem solving' in small, hard-working groups. They are thus encouraged and enabled to produce very reliable knowledge in vast quantities. Their expertise is well respected, and their performative excellence is well rewarded.

---

[14] Ziman, J. (2002). No Man is an Island. *Hermeneutic Philosophy of Science, Van Gogh's Eyes, and God: Essays in Honour of Patrick A. Heelan, S.J.* B. E. Babich. Dordrecht, Kluwer: 203-18.

But society expects and requires of its science just a little bit more than 'normal' achievement. So the question is whether post-academic science provides the conditions in which such ideals can occasionally be made reality. Does it have a few safe niches for the eccentric, idiosyncratic, solitary, awkward individuals who sometimes contribute so enormously to human understanding? Does it give people social space for personal initiative and scientific creativity, and psychological time for fleeting ideas to catch hold and put down roots? Is it hospitable enough to adapt to novelty, and open enough to debate and consider positive criticism. They will need all of these reactions if they are to lead a fulfilling life in science. I don't ask these questions rhetorically, as if the answer were necessarily 'No!' I just think they need to be asked if we are to properly appreciate the place of science in our culture.

NINE

Belief and Action

Knowledge as a guide to action

I may have gone on for too long about life in the Ivory Tower. I spent so many years up there. Its peculiar ways are too beguiling. My excuse is that one could only appreciate its social role if one understood properly how it operated. Surely it is now time to go down to ground level, get outside and see it at work.

OK: so science is a social institution that produces *codified knowledge*. That, we are agreed, is what science is *for*. But it can't be the only source of knowledge in society. True enough, scientific knowledge pervades present-day society. But our way of life is the heir to previous civilisations that got along pretty well with very little science, and the not-so-distant cousin to traditional cultures that had — in some cases still have — practically none of it.

What is more, modern science shares the scene with other institutional sources of knowledge, such as churches and newspapers. It is a monstrous technocratic fantasy that the role of science is to provide society with all of its knowledge, and that these other sources should be ignored or legally suppressed. So we now consider what is *done* with the particular type of knowledge that it produces in such abundance.

Focus, for a moment, on the concept of *knowledge* itself. This is a word with more than a dozen dictionary definitions, hundreds of philosophical renditions and zillions of practical variants. In the large, it enters the life world from science in innumerable forms, through numerous channels, many of them tacit or even without notice. We have already noted the indispensable role of knowledgeable people in public affairs. But let us concentrate on knowledge in its explicit form, as typically codified, communicated and stored in the scientific archives.

In this format, knowledge seems entirely passive. And yet it is connected, by implication, with two modes of action. On the one hand, it is born out of human endeavour, intentionally or serendipitously, by hand or eye or brain, and relates to some features or aspects of human experience, even out to the edges of the observed universe or up to the most abstract patterns of mathematical cogitation. On the other hand, it contains, openly or deeply concealed, the potential for success in the performance of other human enterprises. These may be already planned, only wildly imagined, or entirely unformed. Nevertheless, to know something is not only to have a foothold in the past: it is also to have a handhold on the future. The capacity to enable fruitful *action* — even if no more than the action of getting to know other things — is ultimately what empowers knowledge as a truly social entity.

This is not to proclaim a pragmatic doctrine of meaning. As I have shown at length elsewhere,[1] the whole problematic of the grounds for belief in science is far more complicated than could be encapsulated in a few simple propositions. But it is impossible to make sense of its social role without full recognition of this feature. For its devotees, scientific knowledge often has aesthetic and spiritual qualities; but its societal functions are not primarily decorative or contemplative.

The difficulty is that the verb to 'know' is so absolute. In ordinary usage it is even more downright than its cognate 'truth'. If you say that you 'know' something it implies that you are willing to act on it with complete confidence, as if the contrary could be totally discounted. This is nothing to do with whether what is supposed to be known is actually well-founded, sound, valid, and 'true'. We can see from

---

[1]   Ziman, J. M. (2000). *Real Science: What it is and what it means.* Cambridge, Cambridge UP.

those words that it expresses a positive psychological attitude, with possible future effects, as well as accumulated facts in the present.

If you *truly* know something then in a parallel terminology, you *believe* it completely. Ordinary language has many ways of expressing lower levels of personal certainty, from, say, 'firmly maintaining', through 'assuming' and 'surmising', down to 'having a sneaking suspicion that ...'. But these terms refer to the manner in which individuals hold or communicate various informative statements. In everyday life, these statements themselves are not usually worded in a way that indicates their intrinsic *reliability*. Normally we are left to assess that from the particular circumstances under which they came to be known. A bald statement such as 'The King is dead' could as well be the substance of a wild rumour, the title of a song, or of a BBC broadcast.

Unfortunately, the items of information produced by science usually emerge publicly in just that bald form. The way that they are formulated tends to give the impression that they can be accepted as established 'knowledge', without reservations or qualifications. True enough, the small print reveals that the data are not entirely accurate, the analysis is far from foolproof, and the conclusions somewhat tentative. But these doubts are seldom expressed quite openly, for that would give hostages to fortune, competitors and critics. They are often only apparent to an expert eye, familiar with alternative readings and already primed up with scepticism.

To academic scientists in particular, this public misunderstanding of the status of their findings is disconcerting. As we have seen, they know that credibility is the alpha and the omega of their work. Indeed, it can be argued[2] that the production of knowledge claims which are acceptable to their colleagues is the fundamental purpose of their individual and collective endeavours. The final goal of absolute 'truth' can never be reached. But in their struggles towards

---

[2]   Ziman, J. M. (1978). *Reliable Knowledge: An Exploration of the Grounds for Belief in Science*. Cambridge, Cambridge UP.

it in each particular field of research, scientists get some inti-
mation of how far from it they are. As a result, a significant
part of their professional expertise is to know just how
uncertain their knowledge must surely be—even though
they may be reluctant to proclaim their doubts to 'lay
persons'.

The force of scientific knowledge in society is thus highly
ambivalent. On the one hand, it is held to be the paradigm of
validity and reliability. On the other hand, it is considered
fatally false and flawed. At one extreme, science is said to
produce dinkum, kosher knowledge. Take our word for it.
The world according to science is The World: full stop. We
people in it may have diverse impressions of our condition,
but 'objectively' this is the real McCoy.

It is not only a few gung ho scientists who make these
claims.[3] Many other people nowadays firmly believe them.
A thriving academic industry calling itself 'the philosophy
of science' seems mainly to be devoted to trying to show
how, in principle, they might well be justified. As we have
seen, to admit to being 'unscientific' in practical matters is
considered to be irrational or wayward. So scientific knowledge is a potent weapon, whether for good or evil. By seeking it and acting upon it—or at least claiming its support—
we gain genuine power in the social arena.

At the other extreme, there are those who emphasize its
deficiencies. They bring out into the open its admitted
uncertainties. They point to the dissenters, even amongst
fully qualified professional researchers, to supposedly well-
established findings. Scientific knowledge is presented as
insecure, contrived and overblown.[4] It is said to ignore all
the true values of the human condition. The world according to science is a technocratic nightmare, fit only for
fascists—so they say

It is not just a few religious cranks and political nihilists
who denounce the knowledge claims of science. A growing

---

[3]   Gross, P. and N. Levitt (1994). *Higher Superstition: The Academic Left
and its Quarrels with Science*. Baltimore MD, Johns Hopkins UP.
[4]   Collins, H. and T. Pinch (1993). *The Golem: what everyone should know
about science*. Cambridge, Cambridge UP.

number of quite well-informed people are sceptical of its capabilities. A flourishing scholarly activity, self-styled 'the sociology of scientific knowledge'[5] is busily engaged in 'deconstructing' its most cherished products. All human knowledge is questionable, they say, because it is always culturally 'relative'. And as we have seen, scientific research is often performed under the influence of powerful societal interests. So beware of putting your trust in science, for it will prove a false ally, and deliver you eventually into the hands of an antisocial elite.

One could object, of course, that this is a gross simplification of two perfectly serious points of view. Nevertheless, these are the opposing slogans in a bitter academic conflict. Most sensible people, whether scientifically trained or not, wisely prefer not to take sides in these *'Science Wars'*. But they cannot just exclude science from their social planning. Is there, perhaps, some intermediate position? Should one say, for example, that scientific knowledge is *normally* a *moderately* reliable guide to action, and leave it at that?

We only have to articulate such a statement to see how absurd it is. The unfortunate fact is that one cannot generalise about the status of the knowledge claims produced by organised scientific institutions. Even when entirely authentic, they do not conform to a uniform standard of validity. Sometimes they are as reliable as one could ever wish; sometimes they are really wildly speculative: sometimes they are genuinely disinterested; sometimes they conceal a dangerous bias. On some subjects the scientific beliefs have been confirmed by centuries of successful application; on other subjects they are actually just anybody's guess: in some circumstances you can safely risk your life on them; in other circumstances — 'let the buyer beware!'.

So there is no sense in talking about 'the social role of scientific knowledge' without detailed consideration of the sort of science we have in mind, how it comes on to the

[5]   Collins, H. M., Ed. (1982). *Sociology of Scientific Knowledge: A Source Book*. Bath, Bath UP.

scene, and the part it is assigned in the social drama. This is the theme of the rest of this chapter.

Facts and theories

A common complaint about scientific knowledge is that it is just a load of 'facts'. No scientist should object to that. Reliable facts are hard to come by. They require close attention to detail, alert observation, systematic recording, precise instruments, self-critical honesty and other virtuous or expensive accessories. Indeed, it is something of an accomplishment to succeed in being 'scientific' in any or all of those senses.

Significant facts play a vital role in our society. Think of the dependence of police work, for example, on factual clues, of health care on factual symptoms, of military operations on factual 'intelligence', of engineering on factual measurements. Insignificant facts, assembled as 'data', are the working materials of commerce, industry and government.

Most of the facts of the lifeworld are mundane and self-evident. This makes them rarely contested. But very often what we call a 'fact' depends on scientific research. Think of the immense social influence of factual knowledge of the composition of foodstuffs — the proportions of proteins, fats, carbohydrates, 'minerals', vitamins, in our corn flakes, cheese, hamburgers and coke. Every diet-crazed teenager anxiously scans the packets to be sure to get them right, and so do health and safety officials and executives of food-processing companies. The terms themselves betray their scientific origins. The significance of the information they convey only emerged as a result of research. Debate about them reveals the continuing role of science in determining, refining and redefining them. What is more, such debate is often extremely contentious. It thus reveals the uncertainties, compromises, biases and gaps in the knowledge that they lay claim to.

For scientists, however, the acme of scientific knowledge is not so much factual as conceptual. Facts are just the particulars that are compiled into and represented more generally by *theories*. As we have noted, these are like 'maps'.[6] With their aid we can know something about places where we have not yet been, events we have not yet observed, devices we have not yet made. So their potential for prediction and successful action is incalculable. Do I need to recite their role in engineering, medicine, agriculture, etc.? How would radar have been possible without electromagnetic theory, antibiotics without the germ theory of disease, or the 'Green Revolution' without a basis in genetic theory? Reliable scientific theories clearly perform indispensable functions in our society.

But 'theories' seem so much less tangible, so much more difficult to substantiate, than 'facts'. And yet theories and facts figure inseparably in scientific knowledge. It is not only that science is an *empirical* enterprise. That is, its theories must always be based, however remotely, on discovered, experimental facts. Scientific facts in turn are effectively defined by the theories in which they feature.

Take, for example, a familiar quantitative fact, such as the amount of vitamin A in a carrot. The concept of a vitamin derives from the hypothesis that people or animals presenting certain unhealthy symptoms lacked minute traces of certain substances in their diets. This hypothesis was confirmed by further observations and experiments, and further differentiated and elaborated in a variety of biochemical and biomedical investigations. These were slowly developed into a conceptual scheme whereby the 'quantity of vitamin A' could be linked to an assessment of the health of a laboratory rabbit, or even the title of a chemical reagent.

But these links between the supposed 'fact' and these actual reports of lifeworld events involve many generalisations from experience and inferred causal relationships — i.e., theories. So the 'fact' is not just out there in nature, to be recognised and grasped as a distinct item of information. It

[6]   Ziman, J. M. (1978). *Reliable Knowledge: An Exploration of the Grounds for Belief in Science*. Cambridge, Cambridge UP, Ziman, J. M. (2000). *Real Science: What it is and what it means*. Cambridge, Cambridge UP.

comes to us entangled in a whole network of theoretical concepts, from which it simply cannot be detached.

Don't imagine that I have deliberately chosen a particularly questionable example. The same analysis applies in principle to very hard-nosed sciences such as mechanical engineering. Go out and determine the precise facts about the orbit of a space satellite. You will find them hidden in complex mathematical structures derived from a variety of physical theories, starting with Newton's Law of Universal Gravitation, and you will have to measure them with instruments designed according to optical principles that are even older. Of course the necessary information will include factual statements that are not seriously worth challenging, observations that have been confirmed beyond reasonable dispute, and quantitative data that are exact enough for all practical purposes. But you can't extract *scientific* facts out of these without reference to the scientific 'paradigm' through which they are given meaning.

Nevertheless, as I have remarked, people think of 'facts' as definite and authentic whilst 'theories' are held to be essentially insecure and contrived. We are advised that 'facts are sacred', whilst worldly wisdom cautions us to be suspicious of 'theories'. Scientists learn the art of weaving these contradictory elements together into the whole cloth of a scientific paradigm[7]. But society receives scientific knowledge as a disconcerting mixture of two quite contrary attitudes.

Ways of theorizing

The social role of scientific knowledge thus depends ultimately on the reception of scientific theories. What exactly do these theories tell us, and are they reliable enough to be acted on? This is often dismissed as a philosophical question, but it is actually the central issue in the external sociology of science.

---

[7]   Kuhn, T. S. (1962). *The Structure of Scientific Revolutions*. Chicago IL, U of Chicago Press.

Ideally, we should all be autonomous, intellectually responsible social actors. When scientific theories come into question, we should be able to work them out for ourselves, and make up our own minds whether they are credible. This, I suppose, is the underlying aspiration of the movement for improved 'public understanding' of science.[8] Scientists think that what they know is self-evident, so if only people would listen very attentively to simple, clear explanations of good theories by good scientists then this knowledge would soon become truly public.

Alas, even the best scientific theories are not at all obvious, even when you know all the observational, experimental and theoretical evidence for them. Not even dedicated scholars learn them that way. In practice, students are voluntarily brain-washed into accepting them from their professional mentors in the first instance, and then they come to understand and believe them properly through their own research experience. This is not to say that scientific concepts are genuinely illogical, 'unnatural' or contrary to 'common sense'.[9] It is just that most people have neither the time nor the inclination to appreciate the natural logic and sense in them. So theoretical knowledge has to be taken largely on trust, from accredited authorities.

Some sciences, it is true, seem relatively untheoretical. Their generalisations are primarily *taxonomic* — that is, they offer systematic criteria for the identification and classification of various kinds of natural entities, such as biological organisms. So it could be said that they are essentially factual: 'That is a blackbird. That is a seagull. Note the resemblances — so they both belong to the class *Aves*. But also note the characteristic differences — one belongs to the order *Passeriformes*, whilst other belongs to the order *Charadriiformes*, etc.' Nevertheless, the basic theory that this classification scheme is an outcome of evolution is only now being confirmed by DNA fingerprinting.

[8]   Bodmer, W. (1985). *The Public Understanding of Science*. London, The Royal Society.
[9]   Wolpert, L. (1992). *The Unnatural Nature of Science*. London, Faber & Faber.

Moreover, the social role of this type of scientific knowledge should not be underestimated. Its apparently straightforward theoretical basis makes it an ideal field of popular enthusiasm and endeavour. Amateur, non-instrumental research in 'natural history' not only produces valuable findings. It also opens a door between 'science' and 'society' through which ecological issues can be debated knowledgeably. The political leverage of scientific knowledge in relation to the environment owes a great deal to the way that it is thus brought on to the agenda.

Technoscience, however, is only theoretical by proxy. The knowledge it produces is *practical*. It consists of designs, recipes, treatments, formulae, etc., ready for use. The reliability of these is, of course, of great public concern. Their impact on social practice is of the greatest interest and importance. If they fail to work, or create health hazards, then this knowledge comes into question. It then usually turns out that it is very imperfect. Many modes of failure have not been foreseen, and only a limited number of tests have been made to cover the more predictable ones. The experts reveal the diversity of the alternative designs that might have been followed, and of the uncertainties about their relative merits.

It is all very shocking to a scientist, but not really surprising. Engineering, medicine and other practical arts never pretended to be 'scientifically' reliable. To a technologically alert society, these uncertainties are quite understandable. Institutions for technology assessment, food and drug safety, environmental protection, etc. make it their business to monitor technoscience for risky, inadequately-tested, unreliable knowledge claims, preferably before they are embodied in artefacts or processes. In effect as you will see, they are using science to regulate science.

One could say that the efficacy of *regulatory* science, along with the activities being regulated, ultimately depends on the reliability of more general theoretical knowledge. But this is seldom put in doubt. The disasters to the US Space Shuttle could be explained, and could have been foreseen, without reference to possible deficiencies in the Laws of

Classical Mechanics, or the Chemical Thermodynamics of Combustion. Each was occasioned by a combination of particular circumstances in a particular *context*. What happened was governed by a whole variety of facts and principles, some highly speculative but others quite sound. This is a point to which we shall later return.

It is often argued, on the contrary, that the evident *successes* of technoscience — the great aircraft that fly daily round the world, the epidemic diseases that have been almost eradicated, the smart bombs that can be targetted with such devastating accuracy, the supercomputers that model the climate — prove the truth of its underlying theories. Actually, they only emphasize that even the most theoretical science has to be consistent with empirical reality. This argument doesn't stand up philosophically to the objection that no theory is determined by the facts that support it. As Einstein showed, for example, Newton's theory of gravitation had to be drastically rewritten, and yet remained entirely satisfactory as a guide to the movements of aircraft, missiles and thunderstorms.

But that again is not quite the point I am getting at. Certainly, there is always a remote possibility that our surest scientific mappings of reality are quite mistaken. But the novel general theories of life, the universe and everything that are always being proposed, are very seldom sufficiently coherent or consistent to require these maps to be redrawn. Such theories bring to mind the immortal words of the Duke of Wellington. In response to the greeting, 'Mr Smith, I believe', he replied: 'Sir, if you believe that, you will believe anything!' That is why evolutionary biologists find 'creationists' so impossible to argue with. Having shown that they do not accept the logic of apparently compelling evidence, they can always jump free of any other line of argument that might threaten to tie them down.

Of course, honest intellectual dissent should not be arrogantly dismissed by a complacent elite. And of course there are also occasions, as with the quantum revolution, when the established scientific world picture gets turned upside down. But that happens so seldom and so unexpectedly that we cannot make preparations for it. Certainly, there are extremely academic scientists whose research project is to probe and test our cherished theoretical principles in detail. They often come up with very weird proposals, such as that our universe is 'really' one specimen from an assembly of an infinite variety of possible versions of itself, or that in a zillion years our computers will be so powerful that we shall know everything that ever happened. Like good science fiction, such speculations contribute delightful gobbets of wonder to society. But their scientific reliability eludes public assessment. In the non-instrumental role of Court Jester, science has to remind us that 'the universe is not only queerer than we suppose, but queerer than we *can* suppose'.[10]

The real problem is not with overarching paradigms like evolution and quantum mechanics: it is with theories of the 'middle range'. Take the field of atmospheric physics and start locally. Established theories of the mechanisms producing rain and snow are pretty solid. No doubt there are still innumerable details to work out and many puzzling facts to explain. Somehow, for example, the precise conditions for inducing precipitation by seeding clouds still elude us. But meteorologists use these theories daily to forecast very accurately the sort of weather that will be produced when an air mass of a particular temperature and humidity arrives on the scene.

Zoom out to a whole continent. There again, there is no serious dissent over the main forces governing the system and how they are related theoretically. But the thermodynamics and aerodynamics of the atmosphere are inordinately complicated, and the equations are difficult to solve on this scale. The experts think they have got a handle on them, but they openly admit their uncertainties. Weather forecasts, for example, often state the percentage probability of rain, fog, snow, tornados, etc. in the next 24 hours. Some of these uncertainties are simply observational: the necessary measurements cannot be made on a fine enough

[10]   Haldane, J. B. S. (1927) *Possible Worlds and Other Essays*

scale. Some of them are computational: the equations have to be solved simultaneously at more localities than the machine can deal with. But there are also significant conceptual uncertainties. How should one allow for, say, the effects of vegetation on evaporation, or the role of volcanic dust in the creation of clouds. In other words, here is an extensive body of scientific theory that is far from 'valid' in absolute terms, even though it is remarkably effective in daily practice.

Now stretch the time dimension from a few days to a few millennia. As the mathematicians now tell us, *chaos* rules. The little uncertainties don't precisely average out. Some are magnified unpredictably until they fill the whole screen. So long-range weather forecasts degenerate into rough climatic averages. At the same time, previously insignificant factors, such as slow changes in the concentration of 'greenhouse' gases, begin to exert their influence. But to take account of these requires knowledge of biological processes, such as the rate of growth and decay of forests, or the response of plankton to traces of iron in the oceans, all have to be taken into account. Thus, the necessary theory breaks out of the physical sciences into the life sciences. But no one person knows enough to put all these factors together and estimate their combined effects. In the end, both the input data and the parameters in the equations are so inaccurate that computer simulations of the behaviour of the system are purely notional. So scientific knowledge about climate change is thoroughly bedevilled by missing or imprecise facts, incomplete model calculations, unproven conjectures, and inadequate coordination between experts in different disciplines.

That doesn't mean that the theory of climate change is bad science. On the contrary, it is one of the most effective intellectual enterprises of the late twentieth century. The Intergovernmental Panel on Climate Change — the IPCC — exemplifies post-academic science at its best. Out of a kaleidoscopic mosaic of circumstantial evidence and contestable inferences, an interdisciplinary team of experts have been piecing together a thoroughly convincing picture of the historical run of global climate and its likely future course. Oh, yes, there are a few rejectionists and sceptics amongst the many thousands of scientists competent to assess the situation. And of course the forecast it is much more uncertain than we would like — but no more so than the predictions of the social and economic conditions with which it is now so closely linked.

We have noted how all this started, a century ago, as typically non-instrumental academic research. Now it has put science into the leading role in a global political drama. But note that although climate research is heavily supported by public money, it is still not ordinary technoscience. It is not bureaucratically managed, with well-defined project goals and testable practical outcomes. Nor can the rationale of its findings be made perfectly 'obvious' and 'incontestable' to any single individual. Nobody, not even the most knowledgeable scientific expert, can just go in there and collect together enough of the evidence and counter-evidence to demonstrate conclusively that it is entirely sound — or alternatively, quite unsound.

The IPCC itself does not just summarise the results of innumerable research projects, and average them mechanically. It debates openly to produce a *collective* assessment of the situation. This assessment is not an absolute consensus, but is acceptable to the overwhelming majority of the experts in all the various fields involved. In other words, it is produced by methods that embody the norms and traditions of academic science. It shows just how fruitful these can be in generating socially reliable knowledge.

Nevertheless, the science of *palaeoclimatology* (to give it its official name) is often highly conjectural. Take the 'Snowball Earth' hypothesis. This is the theory that the whole planet, for a certain periods of hundreds of millions of years ago, was gripped by a nearly total Ice Age. There are well-established geological observations that can be interpreted very plausibly in this way. General theories of the atmosphere can be enlisted to show how it might have happened. Perhaps the fragile survival of animal and plant life in a narrow equatorial zone explains an evolutionary bottleneck in that era—and so on. That is just the sort of rather dodgy theory that scientists get a lot of kick out of. Some days they behave as if they believed it, and base their own research on it: other days they suspect that it's a load of nonsense, and deliberately seek evidence to disprove it. Fortunately, the yea or nay of this is of no practical significance. But it is a lovely topic on which to moralise, and illustrates the role of speculative scientific theories as a source of wonder in society at large.

The high point of such theorizing is the *Gaia* concept.[11] From a scientific point of view, this is the idea that the whole planet is something like a living organism. That may be overstating it, but it is now widely accepted that the *biosphere*—the totality of living creatures—interacts so strongly with the atmosphere, the oceans, and the rocky earth that they have always evolved and changed in harmony. Moreover, these changes seem to have actively maintained the surface temperature within the range in which life is possible—never entirely frozen, nor ever a fiery furnace.

What should we make of that? The argument is obviously still very schematic in many details. Gaia's vitality is probably more metaphorical than scientifically meaningful. And it certainly does not imply that the present-day state of the 'Earth System' will automatically survive the destabilising effects of burning off its stores of fossil fuels. On the contrary, the intrusion of global climate studies into the political domain shows that Gaia now has significantly anti-social traits, as well as pro-social ones.

The evocative name Gaia—that old Greek goddess of fertility—has been widely disseminated and appropriated by all manner of non-scientific sects. It means different things to different people. It is vague in outline, and escapes all formal definition. It transgresses the established boundaries between the sciences—physical, biological and

[11] Lovelock, J. E. (1979). *Gaia: A new look at life on the Earth*. Oxford, Oxford UP, Lovelock, J. (1991). *Gaia: The Practical Science of Planetary Medicine*. London, Gaia Books.

human.[12] But that does not mean that it ought to be excluded from 'scientific' knowledge. On the contrary, it just shows that the basic Gaia concept is larger in scale and scope than any which our science had previously envisaged. And without being supernatural, it provides the grandest possible narrative, the widest-ranging 'theory', the most rational rallying point, for the environmental movements that are now so influential in society. What greater social role could science ever play?

Making reliable theories

Theoretical reliability is ultimately what gives scientific knowledge its societal force. Although divorced from technological practice, academic science has acquired an unequalled reputation for just this quality. Actually, as we have seen, its facts and theories are often very uncertain. Nevertheless, they are usually far sounder than any of their competitors. How is this achieved?

One might say that this is a technical question, of concern only to scientists themselves. But as we have seen, scientific knowledge comes 'into society' in forms that often conceal its innate uncertainties. Once these are revealed, the way that the knowledge was acquired becomes of interest to those who aim to use it. They feel a need to understand for themselves how its factual basis was determined and its theoretical framework inferred. This will not free them from dependence on expert authority on just what is to be believed in particular circumstances. Nevertheless, some sense of the general nature of the grounds for such beliefs is an invaluable 'confidence building' measure in deciding whether or not to act upon them.

The folk wisdom is that scientific knowledge is produced by the scientific *method*. This is a collection of foolproof recipes and practical precepts, such as 'measure', 'quantify', 'experiment', 'build a model', etc. Unfortunately, as most philosophers of science now agree, these recipes are all

---

[12]   Ziman, J. (2003). 'Emerging out of nature into history: the plurality of the sciences.' *Phil. Trans. R. Soc. Lond.* A 361: 1617-33.

fallible. What is more, they do not cohere into a consistent system, and ignore many research practices that have proved just as efficacious. As a result, academic discourse on the validity of scientific knowledge tends to be very negative and convoluted. The cagey attitude that this encourages towards mystifying knowledge claims is healthy, although often too dismissive of research results that are simply and flatly counter-intuitive. In practice, however, the formal philosophy of science scarcely figures at all in its public perception and reception.

In previous chapters, however, I have hinted at another approach to this question. We have seen, for example, that academic scientists are driven by the need to prove themselves credible to their colleagues. But their findings must also be novel and original. So they have developed a variety of procedures to make their knowledge claims as convincing as possible to an extremely well informed and sceptical audience. This is especially difficult when these claims are theoretical rather than empirical. So it is worth taking a close look at how scientists actually persuade one another that their contributions to knowledge are sound.

That is a large topic, on which I have written at length elsewhere.[13] It involves social and psychological factors as much as philosophical analysis. In particular, the reliability of the knowledge produced by academic scientists depends vitally on the way they conform to the social norms of their research communities. As we have seen, these norms govern a variety of intellectual and material practices, such as open publication and peer review, which ensure that what gets accepted is logically robust and empirically verifiable. I believe that it is possible to show just how it is that these factors combine and reinforce one another to produce as sound a product as one could ever make by any other means. But this argument is just too complex to be summarised here in a

[13]   Ziman, J. M. (1968). *Public Knowledge: The Social Dimension of Science.* Cambridge, Cambridge UP, Ziman, J. M. (1978). *Reliable Knowledge: An Exploration of the Grounds for Belief in Science.* Cambridge, Cambridge UP, Ziman, J. M. (2000). *Real Science: What it is and what it means.* Cambridge, Cambridge UP.

few words. So let us simply accept it as a piece of practical wisdom — something that we learn from experience to take on trust, scientists no less than other people.

Research tools that shape theories

It is a commonplace that the 'public understanding of science' depends greatly on how it is presented. So it is important to look at the mechanisms by which scientific knowledge is transferred from 'science' into 'society'. But even before this knowledge emerged from the mouths or word-processors of research scientists, it has already been roughly shaped by the tools used to discover it and make it credible. Scientists do not wait until their theories can be put out in a standardised form, like geometrical theorems. Their published findings still bear the marks of the familiar techniques, arguments and strategies. of the research process by which they were revealed and made acceptable to their peers.

Thus, although there is no unique scientific method, scientists employ various characteristic *methodologies* in their research. These are all designed to ensure the maximum possible 'reliability', and yet they are quite diverse in the type of results they actually produce. These differences affect the form in which particular scientific theories are presented to society, and thereby their role therein.

Take, for example, the methodological principle of *measurement*. This often heads the list of typically 'scientific' procedures. Lord Kelvin, that fount of misleading generalisations about science, insisted that it was not possible to know anything scientifically unless it could be expressed numerically. Well, for him all science was just physics, which can almost be defined as the art of the measurable!

Needless to say, good *quantitative* data are easy to replicate, and difficult to dispute. 'If you don't agree, go and measure it yourself'. In high class scientific circles, estimates of the instrumental error are published with the pointer readings, not to indicate any serious doubt but to show just how carefully they have been determined. But of course such numbers mean nothing unless they are inter-related conceptually. Their real virtue is that they can be fed into *mathematical* formulae for almost infinitely elaborate manipulation and analysis.

Thus, the epitome of a scientific theory is a simple mathematical equation—Einstein's equation, $E = mc^2$; or Say's Law, $S = D$; or Heisenberg's Equation, $EW = \mathcal{H} W$; or whatever. Such precision, such logical necessity, such generality, are irresistible. The form in which quantitative theories present themselves implies that they must be entirely believed or entirely rejected. That was evidently what Karl Popper had in mind when he argued that a truly scientific theory had to be potentially 'falsifiable' by a single contrary instance.[14] The theories are even more compelling, though, if they make a successful numerical prediction. So theories like that, if soundly based on empirical data and satisfactorily verified (or at least not seriously disconfirmed) have enormous leverage in society. With their aid people can make bridges, computers, nuclear warheads, and a great many of the other things that typify our culture.

Hold on a moment! Can we always measure the things that we want to talk about scientifically? I will give you that the symbols in Einstein's equation—E for the energy, m for the mass, c for the velocity of light—can all be measured with phenomenal precision, and the quantitative relationship between them verified with exquisite accuracy. So that's a perfectly credible theory, not to be doubted except by a few zany experts. Good for you, young Alfred, and good for us all, too, I suppose.

But what about the quantities being theorised about by Jean Baptiste Say? What he argued was that S, the *Supply* of goods, in a perfect market must always be equal to D, the *Demand* for them. That was at the beginning of the nineteenth century, and only now are economists beginning to drop it from their theoretical armoury. So it has had an enormous social influence. But do those symbols repre-

---

[14] Popper, K. R. (1935 (1959)). *The Logic of Scientific Discovery*. London, Hutchinson.

sent 'quantities'? Could one specify the units in which they were to be measured, and assign plausible numbers to them? Think of a particular example, and you will soon realise how vague and arbitrary they are.

Say's Law comes from the social sciences. But it is no more 'unscientific' in principle than a great many well-founded and widely held theories in the natural sciences. It is easy to see in a general way what he was getting at. Although the result is not immediately obvious, his argument for it makes a lot of sense, and took the best part of two centuries to demolish. And of course economists arrive at other instructive notions by defining similar variables more formally and manipulating them algebraically. But neither Say's Law nor its modern equivalents is the sort of scientific theory that people should be expected to accept unconditionally. Its mathematical form suggests a degree of certainty and quantitative precision that it actually lacks completely.

Theories of this kind are often presented, however, as entirely *abstract*. The terms, propositions or symbolic relationships in which they are stated are not to be identified with actual entities in the real world. They are deemed to belong to a different domain, along with, say, the dimensionless points and perfect triangles of Euclidean geometry and the other mental objects of pure mathematics. There will have to come a moment, naturally, when these pallid creatures of the human imagination are given flesh and blood and made to perform their fascinating tricks on a more mundane stage — perhaps even in the ongoing drama of social life. But this process of *interpretation* is often deferred whilst the theory is refined or elaborated in its abstract form.

Meanwhile, until the practical implications of such theories have been worked out, how should people take the news from scientists about them? Consider Heisenberg's equation, which is central to quantum theory. Apart from the energy, E, the symbols in it do *not* represent directly measurable quantities. $\mathcal{H}$ stands for the 'Hamiltonian', a sophisticated mathematical operation, whilst W, the 'wave function', is impossible in principle to observe. Without a few tough years of higher education, one might get more meaning out of 'Abra Cadabra!'

And yet, in the seventy-five years since quantum theory was formulated, it has had an enormous impact on society. Its technoscientific applications, for example, enabled the invention of computers and lasers, not to mention nuclear weapons. In addition, it is frequently invoked as if it were a general metaphysical principle in aesthetic, theological and moral contexts. Do people believe in it just because it has 'worked'? Or were they well advised, from the beginning, to accept on trust from the academic scientists this extremely abstract 'map' of their discoveries? And how much confidence should they have in the yet more mysterious concepts that are now in vogue amongst scientists, such as strings, branes and parallel universes?

What is significant about theoretical physics, however, is not just that it is mathematically abstruse: it is that it conjures up a whole domain of invisible entities to explain the world that we experience directly. This is typical of many scientific theories. *Nucleons*, for example, are the building blocks of nuclear physics, and are constructed out of *quarks*. Chemical phenomena are explained theoretically in terms of *atoms*. Geology nowadays is a matter of mantle plumes and crustal *plates*. Life is sustained by a restless traffic of highly specific *macro-molecules*. Organisms develop in accordance to instructions coded in their *genes*. And so on. Scientists have invented clever instruments for directly observing these entities, but only after their existence has been inferred from highly circumstantial evidence of an entirely different character.

There is a paradox in the way that theoretical systems of this kind are received into society. On the one hand, their constituents are postulated hypothetically, and often remain so in principle. Thus, for centuries, the atomic theory of matter was a philosophical fantasy. Despite its proven explanatory power in chemistry, many nineteenth-century physical scientists still didn't hold it for real. Even now, when everybody believes in atoms, most ordinary people would have no idea how to demonstrate that they actually exist. As for quarks—well, even a graduate in physicists might be floored by that.

It would be interesting to find out how well the general public believe in another scientific theory that is almost impossible to observe directly—the mobility of the continents. Notoriously, the earth scientists themselves were strangely hesitant to accept this effectively invisible phenomenon.[15] But perhaps ordinary people have more common sense, and are more easily persuaded by the remarkably close fit between the coast lines of Africa and South America that 'there must be something in it'. So the theory of Plate Tectonics is kept going on a slow boil until a new theoretical push convinces the majority of abstract scientists.

On the other hand, some theoretical entities are so heavily promoted scientifically that they acquire tangible reality in the public mind. Scientists have always sought wider understanding of their discoveries. In popularising their theories, however, they tend inevitably to present them in the most elementary terms. The public then get very simplified notions of the nature of their basic constituents.

This is well illustrated by the public reception of the scientific concept of a 'gene'.[16] In the 1860s Grigor Mendel inferred the existence of these hypothetical units of heredity to explain his experiments with peas. Ever since his work was rediscovered at the beginning of the twentieth century, geneticists have been successfully manipulating and mapping 'genes' without any clear idea of their physical nature. Then, in the 1950s, they were discovered to be segments of a chemical compound, DNA, controlling the manufacture of proteins. People now had direct evidence that these were real objects that were passed on, from generation to generation like family heirlooms. But they were also led to believe that every trait of an organism, from the colour of its eyes to

[15] Oreskes, N. (1999). *The Rejection of Continental Drift: Theory and Method in American Earth Science.* Oxford, Oxford University Press.
[16] Rose, S. (1997). *Lifelines: Biology, Freedom, Determinism.* London, Penguin.

the badness of its temper, was hereditable, and was precisely encoded in its genome. It was all too easy then for popular pundits—including many scientists who surely knew better—to make it seem plausible that all human behaviour was determined by our genes, that these had evolved entirely and 'selfishly'. Thus, basic, non-instrumental research produced knowledge of profound significance for our view of our 'human nature'. But because this theoretical knowledge can be made to seem so simple and concrete, it is in serious danger of being taken up by society in such a distorted form that its impact might well prove disastrous.

Assessing evidence

There is nothing new, of course, in the postulation of invisible entities or influences to explain unaccountable events. But has this human propensity been stimulated in our era by the public success of scientific theories involving such secret agents as 'germs', 'genes', 'vitamins', 'prions' and so on? Well, freedom of speech implies that the scientists have no professional monopoly on the promulgation of unlikely hypotheses. These are always being generated by imaginative individuals in all ranks of society. The 'creativity' problem—how do people conceive the germs of a fertile idea in the first place—is endlessly fascinating but shrouded in mystery. For our present purposes it is sufficient that there should be social space and opportunity for this propensity to be exercised.

For working scientists, the problem usually is how to follow up the glimmerings of an idea and to develop it into a rock-solid theory. Philosophers used to say that they have to move from 'the context of discovery' into 'the context of justification'. But even if this context could be defined, it would be open-ended: the process of validation is never finally completed. Scientific theories get into society without absolute certificates of absolute truth. And as we have seen, people nowadays are suspicious of the mere say-so of a club of licensed experts. When a theory seems to be at all questionable, they feel they have a right to be told the nature of the evidence, and on what basis it can claim 'scientific' validity. So the public reception of scientific knowledge is affected by general awareness of the processes by which this knowledge is justified.

Most people are familiar with two standard scientific methodologies. The best-known of these, of course, is *experiment*. Scientists test their theories — frequently to destruction — by observing what happens when they deliberately intervene in the normal flow of events. They contrive unusual circumstances that are particularly sensitive to their theoretical predictions. They explore novel situations on the lookout for deviations from their prior expectations. They construct potentially useful devices and try them out in practice. And so on, in almost infinite variety.

Sometimes — (open the champagne!) — a scientific experiment confirms a remarkable theoretical prediction, or is the site of an unexpected discovery. Mostly, however, its force is negative. Sadly, they agree that this is a theory that won't fly: back again to the drawing board. Strictly speaking, even a good scientific theory is only as firm as the scope of the experiments that have failed to unseat it. And of course, the apparatus is usually elaborately dependent on theory in other ways, so it might not really have been testing that point at all. Experimental research is not just a technical craft. Its rationale is often very subtle.

Nevertheless, this element of 'the scientific method' has penetrated deeply into every corner of modern life. We are advised to take an experimental approach to almost any novel proposal, practical or theoretical. Should the government introduce a new school examination system? The minister reports that she has instructed certain schools to use it 'on an experimental basis' and will in due course report the results to Parliament. These results will, she affirms, be expressed quantitatively, in terms of the number of pupils who eventually get into university, or out of prison, or whatever.

The rationale of systematic experiment is a major non-instrumental contribution of science to the social order. But the general public sometimes takes an over-indulgent view of its outcomes. A theory that seems to have shown positive results in just one (quite possibly faulty) trial is given substantial credibility, whilst other experiments which have not confirmed it are brushed aside. The sceptical precept: 'one swallow does not make a summer' is ignored.

Anyone acquainted with contemporary medical and environmental controversies will know of such incidents. I hesitate to mention specific cases, to avoid being sidetracked into the imbroglios that still surround them. The point is that professional scientists find such reasoning very difficult to counter. This is primarily because its rhetorical effect depends on its being an oversimplified version of a typically 'scientific' approach. 'It was a genuine *experiment* wasn't it? It must have been, because the results were published in a peer-reviewed journal. So how can you say that they must have been faulty?'

The other scientific methodology with which most people are familiar is *statistical association*. The paradigm case is the relationship between smoking and lung cancer. This was not discovered by chance. The relevant data were sought in the first place to test the hypothesis that the one was the cause of the other. They showed such a strong correlation that this hypothesis immediately became highly credible — a genuine, socially significant item of scientific knowledge.

In formal logic, such data are no more (or less) compelling than the results of a contrived experiment. The only difference is that nature has inadvertently set it up for us, and we are clever enough to suss out the results. But that, of course, requires a little luck, and a great deal of professional skill. The logical, methodological and mathematical pitfalls surrounding the collection and interpretation of such data are legendary. Even in the case of smoking, an eminent but unprincipled statistician, Ronald Fisher, was able to earn a handsome and secret fee from the tobacco industry for pointing out that the association between lung cancer and lung disease was no proof of direct causation. It might have been due, perhaps, to the victims having a genetic disposition to both smoking and cancer. Another target for the

Duke of Wellington's robust dictum, I would say; but presumably this weird notion was a psychological crutch for those who were in denial of the harm the practice was doing themselves and others.

Nevertheless, a statistically significant empirical correlation can be good scientific evidence of an otherwise hidden causal relationship. It doesn't *prove* that this relationship is genuine, but makes it much more credible. This is the methodological basis of several major scientific disciplines, such as *epidemiology*. Again, the public frequently take early reports of a small positive effect much more seriously than they warrant scientifically. People quickly begin to speculate about its causes, incidence and implications. Thus, the publication of an unexplained statistical association between autism and MMR vaccination in a relatively small sample of cases caused a great public uproar. But attention focussed on theories about how this dangerous phenomenon could come about and what should be done about it. Nobody seemed to take any notice of the results of larger, more systematic surveys that failed to confirm that there was any effect at all.

The response of the general public to an item of scientific information depends, of course, on its potential threat to human welfare. We have already noted, and will return to, the *risk* dimension in the relationship between technoscience and society. Some hazards, such as skiing or train travel, are so tangible that there is no arguing with the accident statistics. But when a risk is hypothetical — for example, the effects of genetically modified foodstuffs on human health — its scientific credibility becomes a matter of direct public concern. People feel they have a right to assess this by their own criteria. They go back to the factual evidence for the theory at issue and judge it according to relatively naïve popular conceptions of scientific validity.

These have been just some examples of situations where people claim that they have a legitimate voice in the 'context of justification'. In effect, when 'society' comes into close contact with 'science', the boundary between them is easily transgressed. It is not just that scientific knowledge percolates into popular consciousness and influences social action. Social values and beliefs travel across the boundary in the other direction, and intrude into research practice. In an open society, these counter-currents become reflexive. Scientists begin to expect a say in the way in which their knowledge will be used. Conversely, people want a role in the process by which this knowledge was produced. Even though the mutual understanding is often imperfect and ill informed, it is surely beneficial to the social order.

Interpreting theories

As we have seen, to *believe* a scientific theory is to be ready to base ones actions on it. To be of any use, it must therefore provide reliable information about things one does not already know. That implies *prediction*, not only of events that have not yet actually taken place, but of what might be disclosed in the future about current or past events of which one happens to be ignorant.

Needless to say, a scientific hypothesis can be made much more credible by a successful prediction, especially when what is predicted seems otherwise very unlikely. But that is not, as is popularly supposed, the keynote of scientific practice. Or, rather, if it were such an unlikely keynote, it would require a much wider, more liberal interpretation of the term than the customary usage. In many cases—Darwin's theory of the origin of species, for example—a simple, unforced explanation of a great many known but apparently unrelated facts is just as compelling.[17] Here again, professional scientists are often hard put to it to counter public misconceptions of the nature of scientific argumentation.

In practice, however, the social role of scientific knowledge is continually hampered by quite a different problem —how should theories be *interpreted*? A theory is a *map*, an assemblage of *abstractions*, a *representation*. For people to 'believe' in it they have to be able to bring it to bear on their lifeworld, taken-for-granted actions. So what do its mysteri-

[17] Ziman, J. M. (2000). *Real Science: What it is and what it means.* Cambridge, Cambridge UP.

ous terms, variables, symbols, entities, operations, actually *mean*? What do they relate to in the 'real world'? This is not simply a very vexatious philosophical question. It is an ever-present issue in the day-to-day interactions of science and society.

True, this issue seldom presents itself directly in the normal practice of technoscience. Basic theories there are hidden away in the back room, not in the front office. The task of interpreting them is performed inside the system, in the 'development' phases of 'R&D'. The knowledge that technoscience supplies to our society is already engineered into practical artefacts, technical recipes, therapeutic procedures, instructions to users, solutions to actual problems and such like. These may be weirdly elaborate, and maddeningly obscure. The services of an expert may be required to extract from them the desired information. But it is only in exceptional circumstances — the global tragedy of AIDS might be an example — that the underlying theory of an established technological practice comes under public scrutiny.

Academic scientists, on the other hand, often become so familiar with the theory of their specialty that they do not know how to interpret it in other terms. They have forgotten — or were never properly taught — how to relate their particular scientific domain to the uncertain domain of everyday life.[18] Their conjectures, predictions, discoveries and explanations are concerned solely with entities inside this imagined world. There is no occasion for their thoughts to stray outside it. They become so immersed in it that they even have difficulty in translating their concepts into the language of a neighbouring specialty.[19] Cosmic strings, gravity waves, quarks, genes, species, social actors, totems, liquidity preferences, or whatever they may be — intangibly abstract or provocatively material — these are the constituents of their peculiar reality.

[18] Solomon, J. (1992). *Getting to Know about Energy in School and in Society*. London, Falmer Press.
[19] Galison, P. (1997). *Image and Logic: A Material culture of Microphysics*. Chicago IL, University of Chicago Press.

So most academic scientists have great difficulty in presenting their theories to the public, especially when these are still disputed by their research communities. It is like trying to explain a family quarrel to outsiders, and it is often a considerable achievement to give other people intelligible, if stereotyped character sketches of the principal actors. (But do they realise that righteous Uncle Arthur is actually a bit shifty, conventional Aunt Dolly occasionally behaves very oddly, and kind Cousin Ethel is a covert bully?) As we have seen, the popularised version of a scientific theory is seldom a reliable guide to action. The entities that figure in it are depicted as if they were as real as tables and chairs. In truth, they are abstractions whose properties have necessarily been simplified to fit into the theoretical scheme in which they appear.

What is worrying is that even the scientists who know all these detailed complications and uncertainties are usually unaware of the innate limitations of their theories. The fact is that a scientific theory, like any other 'map', can only represent one aspect of many-sided, many-splendoured reality. This aspect is often selected primarily because it is easy to depict, or schematise. One of the most efficient strategies of research, for example, is to dissect a compound object into separate components, and study these as if they were independent of one another. To understand molecules, say, discover the properties of atoms, and so on.

Needless to say, this policy of *reductionism* is the source of most of what is known to science. But it is immediately obvious that this knowledge is likely to omit the *holistic* features that had to be discarded in the dissection operation. A molecule is not just an assemblage of atoms, but a *structure*, with properties greater than those of its parts. Although many philosophers and scientists seem to be unconcerned about this objection, it comes quite naturally to any intelligent person. Indeed, it is frequently (and properly) voiced by public critics of scientific knowledge, even in its technical applications. They say, for example, that scientific medicine treats human beings like machines, as if their organs could be disassembled and worked upon separately without any regard for the 'whole person'.

Because reduction is such a powerful research methodology, this critique applies to many established scientific theories. So there is a significant limiting factor in the relationships between science and society. But it only reminds us that current scientific theories do not include everything we presently know or would like to know. The fact that they clearly have limits does not, of course, guarantee the reliability of just any of the 'alternative' belief systems that might be proposed to supplement or supplant them. On the other hand, it does not rule out the possibility of developing a better scientific understanding of the overall properties of compound entities with many complex constituents.

Nevertheless, the knowledge produced by present-day research methodologies is typically reductive. It represents the world as a 'nested hierarchy of systems' — that is, it postulates a succession of 'levels', in each of which the constituents are compounded of the constituents of the level below. For example, *societies* are compounded of *organisms*, which are compounded of *cells*, which are compounded of *molecules...*, ... etc.

Let us not go into the reasons why this mode of analysis seems so natural and effective.[20] The point here is that it is relatively easy to present, level by level, and make publicly intelligible. In effect, the theory of what takes place at each level is a *system* of relatively distinct entities, with more or less separable properties and interactions. As we have seen, this type of theory is easily simplified for popular consumption. Indeed, in the case of 'genetic psychology' it has penetrated so deeply into our culture that it has become the basis of a new 'folk science'.

What this reductive analysis also facilitates is the development of theoretical *models*. I spare the reader any discussion of the deeper philosophical significance of this characteristically scientific notion. All I am saying is that if a

---

[20]   Ziman, J. (2003). Emerging out of nature into history: the plurality of the sciences. *Phil. Trans. R. Soc. Lond. A* 361: 1617-33.

system is not too 'complex' (a term that is more subtle than it seems at first sight) its behaviour can be calculated, or computed, or otherwise prognosticated. The mental 'map' springs to life as an instrument for rational *prediction*. And since that is precisely the active social function of a scientific theory, we may have hit the bull's eye.

Thus, scientific knowledge often enters society in this form. A scientific theory is seldom easy to understand in its original technical language. So it seems helpful to translate it into a model, almost as if ready to be put into use. The need for expert interpretation is avoided. With sufficient computer power, people can work out its implications for themselves. The role of science in society is thus notably enhanced.

That is only true up to a point. Many of the benefits (and dis-benefits) of technoscience, are empowered by theoretical modelling and computer simulation. But this has its limits. The mathematical precision of the laws of physics and chemistry enables the design of novel engineering structures and materials, although even these sometimes show unforeseen flaws. Reasonably accurate weather forecasts a few days ahead are another routine achievement, even though these are inevitably randomised by chaos in the longer term. But scientifically modelled foresight begins to fail when biological factors enter the equations — for example, in the effort to create 'designer drugs' or to treat specific ailments. And when the constituents of the model are conscious human actors — in economics, for example, or in nuclear weapon strategy — the deficiencies of this approach become very obvious. The scientists who make and run such models learn something from their behaviour, but only very indirectly about the aspects of the real world which they are supposed to represent.

What I am getting at, once again, is that science seldom produces theoretical knowledge in a socially digestible form. Even its pre-cooked dishes, which seem to go down so well, should be taken with more than a pinch of worldly-wise salt! The interpretation of scientific theories for public use is not something that should be left to a thin stratum of popularisers and pundits who think they are mediating between 'science' and 'society'. As with all knowledge transfers between distinct cultures, it is quite a 'thick' process requiring both technical expertise and common sense.[21]

Problems of interpretation also stretch much further back into the research culture than is usually realised. Genuine scientific progress is never easy. Scientists typically try to make it less difficult by tacitly simplifying their representations and models. The limited scope of many of their basic assumptions can lead to serious misconceptions about the real significance of their findings. The unmasking of these misconceptions is one of the non-instrumental contributions of academic science to society.

Fitting the context

The knowledge claims of academic science are typically *generalised*. They have to be acceptable to a potentially universal community. Of course the processes by which they are generated — the observations, experiments, calculations, 'brain waves', etc. — take place at specific points of time and space and under quite varied social conditions. So do all human activities, don't they? The art of research is to infer from these particulars a representation that is independent of them. Even when, as so often in the Earth Sciences, they are field data from a specific geographical locality, their scientific significance is their place in a larger theoretical picture.

But the settings where this knowledge performs its social roles are almost always highly *localised*. That means more than their space-time coordinates. Scientific knowledge is brought into play in specific *contexts*. These vary greatly in their material and social circumstances. Much depends on the abilities and goals of the relevant human actors, the instrumental resources available to them, the institutional frame and the other knowledge systems operating in the

---

[21]   Geertz, C. (1973). *The Interpretation of Cultures*. New York NY, Basic Books.

neighbourhood. Differences in any of these contextual features can profoundly affect the way in which an item of scientific information is actually received.

Once again, this is not a significant issue for techno-science. 'Mode 2' problems are conceived in 'contexts of application' and that is also where their solutions are eventually going to be put to use. A vast amount of technical work — generalising, theorising, experimenting, designing, developing, testing, etc. — may have to intervene before the appropriate artefacts are put on the market. But the organisations that carry out this chain of operations are conscious of its ultimate purpose, and shape their products accordingly. If these products don't fit their 'contexts of application', then they don't sell, or patients are not cured, or terrible accidents occur, or the army is defeated, or something similar. Of course, since we are none of us perfect, such disasters may often happen, in which case it's back again to the drawing board again.

So translating and/or reshaping scientific knowledge to fit its non-scientific context is normal practice in the commercial sector. One of the features of modern technoscience is that this process is usually highly institutionalised. The marketing divisions of big industrial firms don't just advertise and sell their wares. They provide the public with customer advice, after-sales service, and other forms of technical information. What is more, through the quasi-scientific medium of consumer research, they enable 'the context to speak back'.[22] Indeed, failure to keep such channels open can be very costly. Thus governments frequently spend large sums on elaborate scientific initiatives to support their industry or agriculture. All too often, this would-be instrumental research is quite wasted because it is carried out in isolation from its context of application.

For 'Mode 1' researchers in post-academic science, however, this is normally not a matter of direct concern. As we have seen, it is hard enough to persuade them to translate

[22] Nowotny, H., P. Scott, et al. (2001). *Re-Thinking Science: Knowledge and the Public in an Age of Uncertainty*. Cambridge, Polity Press.

their theoretical findings into general knowledge that can be grasped by non-scientists. They seldom have occasion to transform this knowledge into a recipe for action in any particular context. And when this is demanded of them, they find it extraordinarily difficult. Their whole mind-set is against it. Precisely because their whole effort is directed at inferring generalities from specific, local events, they cannot easily engineer their thinking in reverse.

Academic experts are also notoriously specialised. They are trained to perceive only certain aspects of a local situation, and to bracket out everything else with the formula 'other things being equal'. Being unaccustomed to grasp things whole, they enact the fable of the three professors who encountered an elephant in the dark. ''Tis a tree!' exclaimed the one who blundered into a leg—for he was a botanist. 'Nay, it's a monstrous snake!', cried the herpetologist, caught up by the trunk. 'Surely, it is a mounted lancer!' moaned the unfortunate military historian, pierced by a tusk.

Actual social contexts are not textbook problems. They are not designed to exemplify the principles of a particular scholarly discipline. Even when they are not perplexingly multidisciplinary, they are usually much messier than their laboratory counterparts. Reality refuses to conform to the simplified circumstances assumed in the theoretical models. When a genuine elephant slides down a genuine plank, its bulk and vitality confound its 'physics representation' as a point mass. What is more, it will be linked to its 'context' by restraining chains and goading mahouts. What are the dynamical correlates of the rage aroused by its capture, or of the appetite stimulated by the enticing smell of food?

To use the language of mathematical theory, even when the model is quite simple in principle, the 'boundary conditions' are always complicated. The 'context' refuses to stay fixed and changes in response to other external circumstances. So the expert is faced with a sequence of events that cannot be accounted for without knowledge of their historical antecedents, current effects and supposed outcomes. That is not to say that these events are scientifically inexplicable. But that their explanation can no longer be expressed in the timeless generalities of an abstract theory.

Metaphorically speaking, the task of the expert adviser is to consult his or her knowledge 'map', and compile an 'itinerary' that includes specific information of what has happened or might be expected to happen at each stage of the journey. Some of this information may be highly *contingent* — that is, unique to the particular circumstances. Real life does not run on tram lines. The unfolding of events is *path dependent*, even for the steamrollers of economic science.[23] So the uncertainties of the map are compounded by the happenstance of events.

Suppose, for example, that I have a mind to grow a few acres of trees, as a minuscule mitigation of global warming. The general theory of climate change supports this action in principle, but will it really have any effect? So I look more deeply into the theory, and discover that the amount of atmospheric carbon that will thus be segregated depends on the species of tree, on the local climate, the type of soil, its microfauna, the previous crops on the land, the proposed mode of cultivation, what happens to the wood when it is finally harvested — even on the economic prosperity of whoever inherits my property. I cannot possibly discover and digest all the relevant scientific knowledge on my own, so I consult various experts. These will need to be informed in detail about the history and current condition of the site, and my plans for it. And what I seek from them will be an account of the likely consequences of the various courses of action that I might decide to take — a rational conception of the shape of things to come in this locality. Shall I achieve more, in the end, with slow-growing oaks than with fast-growing poplars — and so on?

In other words, the requisite knowledge typically takes the form of a *narrative*. A sequence of scenes has to be sketched out, passing from a partially known past into a hypothetical future. Like any good story, it will include

---

[23] Arthur, W. B. (1994). *Increasing Returns and Path Dependency in the Economy*. Ann Arbor MI, U of Michigan Press.

unforeseen events that have already taken place, and will make allowance for their possible recurrence. Sometimes a vivid 'picture' will convey far more than words. Indeed, several alternative 'scenarios' may be required, to cover the risks and uncertainties inherent in the situation. As the movement for 'technology assessment' has shown, the narratives offered to consumers by technoscience are always dangerously oversimplified. Reliable scientific advice on complex issues cannot be made as clear and unequivocal as the instructions for assembling a prefabricated bookcase.

In effect, an academic scientist seeking to 'contextualise' his or her expertise is faced with a daunting task. By what rational criteria can one extrapolate general science facts into particular science fictions? What will support ones mind and reputation as one is parachuted out of the scientific domain into some booming, buzzing locality in the surroundings? What language do the natives speak? Are they friendly?

Fortunately, there is an old-established solution to this problem. Wherever a cultural gap becomes an obstacle to desired action, intermediaries flow in to fill and bridge it. This is as true of knowledge cultures as it is of ethnic or commercial ones.[24] In the present case, this bridging function is performed by expert *practitioners*. On the one hand, as we have seen, their training gives them access to current scientific theory. On the other hand, their day-to-day employment is highly localised and contextual. The job of the physician, the architect, the civil engineer, the planning consultant, the legal adviser, *et al.*, is to identify and download from a body of generalised knowledge whatever is required in each particular case and to transform it into a therapy, a design, a recipe, a brief, or other instructive narrative for the benefit of his or her customer.

Of course these professions fulfill many other functions in modern society. But I am only concerned here with their

[24]   Galison, P. (1997). *Image and Logic: A Material culture of Microphysics*. Chicago IL, University of Chicago Press.

official role as knowledge brokers between science and society. Needless to say, that is a very elementary point. And yet it raises many quite basic questions. Are their practices efficacious? Do they normally do a satisfactory job? Are they answerable intellectually to 'science' or to 'society'? Who pays them? How do they sustain their moral credibility? Are there other areas of society and/or fields of knowledge where they might be needed? Do we need new institutions to augment or replace them?

These are only some of the questions that can be asked about the place of professional experts in modern society. But they are essentially *sociological* questions, so I shall defer them until the next chapter. In effect, I accept the conventional view that independent practitioners are not part of the machinery for the *production* of knowledge. But they are not just 'ordinary' elements of society. 'Is there a doctor in the house?' is a dramatic appeal full of practical implications. Their highly specialised work, individually and institutionally, is essential to the relationship between science and society. It is true, as we have seen, that academic scientists are often called in to mediate this relationship. But that is a feature of our social order, not of our knowledge of the world.

Research in the context of implication

To the experienced practitioner, the local narratives through which scientific knowledge typically enters the taken-for-granted knowledge of the lifeworld are never just past history. They are guides into the future. Each of them indicates a possible outcome to present action. 'I used that drug in a case rather like this several years ago, and it didn't have much effect: perhaps we should try another brand'. And of course, it is towards desirable outcomes that expertise is applied. In other words, the *implication* of the use of the knowledge is central to its meaning.

This consideration is reflected back into the knowledge production process. The 'context of application' is transformed into a *context of implication*.[25] Research projects must not only take into account their potential for practical use. They are planned in relation to the likely response of possible users and the further consequences for them and others. Purely technical promise is not enough. Various scenarios of application have to be envisaged and evaluated before the work begins.

In commercial technoscience this is already the normal practice. A pharmaceutical company obviously does not undertake research and development on a drug to treat a specific disease without investigating its symptoms, incidence and market prospects. This includes the circumstances of those who suffer from it, the side effects of the treatment and its long-term promise to them. The research programme is as much shaped by its possible social and psychological consequences as by its more tangible medical potentialities.

That is why even very sophisticated technoscience is never as 'scientific' as it claims. In retrospect, the implications of a research project are what matters. And yet they are far less predictable than its hoped-for immediate applications. The context of implication for the Manhattan Project, for example, was dominated and limited by the desire to defeat Nazi Germany and Japan. Only a few visionaries perceived that it ought to have included at least a hint of the Cold War and other features of the post-Hiroshima era.

And yet, as science becomes more closely entangled with society, the context of implication grows in importance. Indeed, in the human sciences it both supersedes or incorporates the context of application. Research in sociology, for example, is seldom directed towards the solution of specific problems or the development of practical 'innovations'. Its purpose is usually to understand what is going on, and to provide people with a better appreciation of how to achieve

---

[25]   Nowotny, H., P. Scott, et al. (2001). *Re-Thinking Science: Knowledge and the Public in an Age of Uncertainty*. Cambridge, Polity Press.

their goals.[26] Or it may be melded with other social processes into 'action research', where the intentions of the actors are closely observed in relation to the actual consequences.

Card-carrying technocrats dream that all the uncertainties surrounding the application of scientific knowledge can be 'fixed' by a superior scientific strategy. They believe in the feasibility of constructing formal 'models' of society, or of individual behaviour, which ought to give reliable advice on what will happen if a research project is successful. They would be wiser to realise that the consequences of any significant social event are determined (if that is the right word for any historical process!) by numerous intangible 'local' circumstances.

In particular, the way in which knowledge is received and applied depends greatly on the values, norms, preferences and prejudices of the actual people involved. Information about these is not to be found in the scientific archives or on the Internet. It has to come directly from 'the horse's mouth'. In a free society, the outcome of any action can only be justly assessed by the people most likely to be affected by it. Even a research project cannot be made to fit its context of implication without the active help from those deeply involved in formulating it.

[26] Lindblom, C. E. and D. K. Cohen (1979). *Usable Knowledge: Social Science and Social Problem Solving.* New Haven CT, Yale UP, Lindblom, C. E. (1990). *Inquiry and Change: The Troubled Attempt to Understand and Shape Society.* New Haven & London, Yale University Press.

TEN

The Pluralism of Global Knowledge

Micro-narratives in wider picture

As we have seen scientific knowledge mainly enters social practice in the form of '[a] multitude of many micro-narratives embedded in local and particularistic contexts'.[1] In most cases this knowledge is plainly instrumental in its origins and use. Technoscience typically produces many technical stories — some true, some mere fables — along with its artifacts, and its instruction manuals. Every teenager gets to know, for example, that radio waves penetrate brick walls: otherwise, how would their precious mobile phones work indoors? But what a variety of urban myths this jolly invention must already have spawned.

In other cases, items of knowledge are retrieved from the academic archives and translated into lifeworld terms because of their obvious utility. For example, theories of climate change are transformed into stories explaining unusual weather events, and widely reported as such in the news media. Sometimes, of course, as we have seen in the case of genetics, the supposedly 'scientific' account of a real-life situation is oversimplified and misleading. Sometimes again, especially in the human sciences, one side of a highly contentious theoretical issue is so loudly trumpeted that it is taken to be just 'common sense'. Nevertheless, numerous fragments of quite sophisticated scientific

---

[1] Nowotny, H., P. Scott, et al. (2001). *Re-Thinking Science: Knowledge and the Public in an Age of Uncertainty.* Cambridge, Polity Press.

knowledge do get safely into the public domain, particularly if they can be made to appear 'relevant' to everyday life.

Above all, however, there remains an abiding interest in the wider picture. People feel that they ought to be told about 'the world according to science'. As we have seen, they are always hungry for objects of curiosity and wonder. This traditional non-instrumental role is still being performed in the post-academic era. Indeed, lavish military support for such technoscientific enterprises as ballistic missiles, surveillance satellites, seismic detector networks and anti-submarine defence systems spill over into unparalleled facilities for the peaceful exploration of the heavens, the earth and the deeps of the ocean.

What sort of 'world picture' are people actually getting? That's impossible to say. The public sources of general scientific knowledge—schools, universities, popular science journals, books, newspapers, radio and TV programmes, etc.—produce a great mass of information on an immense variety of topics. But these are only the individual pixels in an image that is too large and multidimensional to grasp as a whole. As we have seen, the professional philosophers are mostly entangled in formal minutiae, whilst we amateur philosophisers mostly emanate clouds of waffle which lend the world pictures a host of possible backgrounds.

Nevertheless, it is customary to refer to 'science' as a more or less coherent enterprise with a reasonably self-consistent message. In this usage, of course, the focus of attention is narrowed so as not to range too far over the border from the 'natural sciences' into the 'human sciences'. But within these limits, the principal claim of science is that it produces knowledge of 'objective reality'. Certainly, this is what scientists themselves largely believe, and the general public does not seriously doubt.

In principle, as we have seen, scientific knowledge is as distinct from 'reality' as a map is from the landscape that it represents. But this is a metaphysical distinction which applies generally to everything we say to one another about our experiences, however mundane. For all practical (and

almost all impractical) purposes, this knowledge is genuinely and reliably about the world we all happen to share,
however mysterious that world sometimes seems to be.

Just how science produces knowledge with these properties is another matter, which I have discussed at length elsewhere.[2] Nor need we discuss further the uncertainties and
mutations that are inherent in the evolutionary processes by
which all such cultural entities are generated.[3] In any case,
apart from relatively localised 'revolutions', what has been
called the *epistemic core* of science[4] has grown steadily over
the centuries. Sociological relativists and other sophists
would have it that this core is actually empty, or that it is an
ideological construct of a powerful elite. Well that sort of
argument, for whatever it is worth in academic circles, cuts
no ice with the general public. So far as they are concerned,
scientists don't always get things right, but when they do it
is 'nothing but the truth' as the jury oath goes.

But the other half of the legal catchphrase — that science is
'the whole truth' — is losing public support. Some scientists
still proclaim this doctrine,[5] but they are hard put to it to
defend it philosophically. People believe that there are
bodies of genuine knowledge about matters on which science
is silent, or ill informed — and they act accordingly.

This is clearly an extremely important limiting feature of
the role of science in society. It would take us too far afield to
discuss the nature, quality, claims and capabilities of these
competing knowledge systems. But we do need to clarify
the actual extent of the legitimate domain of science itself. In

[2]   Ziman, J. M. (2000). *Real Science: What it is and what it means.*
      Cambridge, Cambridge UP.
[3]   Ziman, J. M., Ed. (2000). *Technological Innovation as an Evolutionary
      Process.* Cambridge, Cambridge UP, Wheeler, M., J. Ziman, et al., Eds.
      (2002). *The Evolution of Cultural Entities.* London/Oxford, British
      Academy/Oxford University Press.
[4]   Nowotny, H., P. Scott, et al. (2001). *Re-Thinking Science: Knowledge and
      the Public in an Age of Uncertainty.* Cambridge, Polity Press.
[5]   Gross, P. and N. Levitt (1994). *Higher Superstition: The Academic Left
      and its Quarrels with Science.* Baltimore MD, Johns Hopkins UP, Gross,
      P. M., N. Levitt, et al., Eds. (1996). *The Flight from Reason.* New York
      NY, New York Academy of Sciences.

what sense does it not tell 'the whole truth' about life, the universe, etc? And what general tools of thought does it now provide for those of us who seek to improve the lifeworld into which we have happened to be born?

The empires of modernism

To put it simply: the future of science as a system of knowledge can no longer be envisaged as an indefinite extrapolation of the 'modern' world view. The 'scientific' perspective that has dominated 'Western' natural philosophy since the seventeenth century is now losing focus. Science can no longer be expected to zoom in and sharpen up its image of every imaginable feature of the world. This is not a result of some dreadful defeat at the hands of anti-scientific, post-modern, constructivist sociological relativists. It arises out of the very success of the scientific enterprise in understanding better its own capabilities and limitations.

This argument can be structured around a 'geographical' metaphor. In an abstract sense, the 'sphere of knowledge' itself is also being 'globalized'. In academic discourse we refer glibly to 'fields' and 'domains' of knowledge, to 'unexplored areas' and 'breakthroughs', to 'frontiers' and 'boundaries'. As we have seen, the notion that scientific knowledge can be 'mapped' out is intuitively meaningful, although still almost impossible to explicate in other terms[6].

In effect, our overall map of the 'scientific domain' is becoming more 'global' in much the same sense as our overall geopolitical map is also doing so. But that does not mean that it is getting nearer to becoming 'universal'. Our vision of that sacred goal is, if anything, receding further into the mists of impossibility.

To understand what has happened, we need a historical perspective. From its beginnings, 'modern' science has fostered a *mechanistic* view of the world. This is usually

---

[6]   Toulmin, S. *Human Understanding*, Vol. 1, Oxford: Oxford University Press, 1972; Ziman, J.M. *Reliable Knowledge: An Exploration of the Grounds for Belief in Science*, Cambridge: Cambridge University Press, 1978; Ziman, J.M., *loc. cit.* 2000.

attributed to the Newtonian synthesis of what we now call physics, and its subsequent elaborations. But the other branches of the natural sciences that emerged more or less independently were activated by a somewhat similar philosophy. By the middle of the nineteenth century these branches of the science tree had established themselves as distinct *disciplines*.

In each discipline, specialist researchers learnt to study nature in accord with a distinctive *paradigm*. That is to say, the scientific map consisted mainly of a few well-explored areas, labelled 'physics', 'chemistry', 'botany', 'zoology', and so on. These were disconnected islands of scientific knowledge scattered over an uncharted ocean. Or perhaps they were isolated clearings in a dark, densely forested continent. Elsewhere, however, scientific ignorance reigned.

The various paradigms that governed these islands were seldom literally 'mechanistic'. But they were optimistically simplistic. Scientists in each discipline tended to focus on the restricted range of phenomena that they seemed to be able to explain quite nicely with a limited tool kit of concepts. What a wonderful lot of physics you could do with billiard-ball atoms, classical mechanics, gravitation and electromagnetism. All that you needed for chemistry was the periodic table and space-directed valence bonds. Biology was a bit messy, but provided you didn't ask too many questions about 'protoplasm' you could build up organisms out of cells, and crank an evolutionary handle to understand their diversity and the way they functioned.

Genuine scientific knowledge was thus supposed to be limited to the core ideas of a relatively small number of disconnected paradigms. Within each discipline, these ideas could be represented in terms of simple 'laws', quasi-mechanical models and well-ordered classifications. Indeed, the causal mechanisms operating in each core set were so obvious that the outcome of every action could conceivably be predicted—that is, said to be 'determined' in advance. Moreover, some of these models looked so straightforward that they could be broken down mentally into their constituents and then reassembled to perform just as before. The notion that the behaviour of a compound system could be 'reduced' to the properties of its components was elevated into a general metaphysical principle.

In Victorian times, the differentiation of scientific knowledge divided it into disciplinary areas that harmonised with its political and cultural 'modernism'. In effect, each discipline was a well-ordered nation-state of the mind, governed rationally by a strong paradigm. Beyond its frontiers, there was only erratic, irrational barbarism. Scientific progress was henceforth to be devoted primarily to the expansion of disciplinary authority into previously unsubdued regions of the natural world.

In practice, these expanding empires sometimes collided. The chemical atom was enlarged to include the electron, which was also claimed as an elementary particle by physics. The biological cell was surveyed internally in finer and finer detail until the life scientists stumbled through the forests of proteins and nucleotides into chemistry. From time to time an undefended territory might be colonised from an established discipline, as ecology was from systematic biology. New disciplines such as evolutionary genetics might also emerge in completely unexplored domains. And of course, even an apparently stable disciplinary regime might be 'revolutionised' by a radical change of paradigm, as classical physics was by quantum theory and relativity.

The geopolitical metaphor thus shows how well the world view of 'modern' science conformed with the other general characteristics of the 'modern' age. Nineteenth and twentieth century scientists looked forward to further expansions of scientific rationality, until the disciplines had more or less joined up to form something like a coherent map of scientific knowledge. They did not worry overmuch about whether this map would ever be complete. They presumed that there would almost always be an 'endless frontier' across which intrepid explorers would be tempted to cross, so as to make further discoveries. Only at the very end, when there were no more 'irrational' lands to conquer, would the disciplines disappear. They would all have been 'reduced' to satrapies of a world empire of 'universal' science, where a 'theory of everything' ruled.

Modernism works itself out of its job

Unfortunately, the simple, mechanistic models of the nineteenth century did not prove equal to their imperial mission. Since then the scientific map has indeed been immensely enlarged. There remain few completely blank spaces. The territories between the traditional disciplinary localities are now reasonably well understood. For example, physics has become fully connected with chemistry, and molecular biology now covers the area between these disciplines and biology. New sciences extend into the earth, the biosphere, the atmosphere, the solar system and the cosmos. Scientific knowledge stretches far back into the past — and is expected to do so into the future.

But by their competitive search for novelty, modern scientific endeavours, like modern capitalist enterprises, inevitably foster the seeds of their own supersession. Continually probing, innovating and elaborating, they range ever more widely, and reveal a much more complex and uncertain reality than was previously imagined. By the end of the nineteenth century science had already encountered numerous entities and phenomena, ranging from volcanic eruptions to outbursts of rage, that could not be made to fit into the established schemes.

Indeed, almost everywhere we look, we now find burgeoning diversity. Physicists talk of their 'zoo' of elementary particles. There seems no end to the number of different types of biological organisms. Even the astronomical universe, which used to be just an assembly of spherical stars and spiral galaxies, is full of weird objects, spinning giddily, exploding violently, emitting every sort of radiation, spewing out beams of energetic particles or gulping matter into its black depths.

Again, the scientific imagination is beggared by the sheer complexity of so many natural systems. It seems that many so-called 'elementary' particles are composed of whole families of quarks. The 'protoplasm' within a cell turns out to be as finely structured, spatially and dynamically, as a whole multicellular organism. Living creatures are kept going by interlaced reaction cycles involving innumerable complex molecular species. Ecosystems are both diverse and complicated. The atmosphere exhibits irregular patterns of turbulence on every scale. And so on.

Needless to say, a century of research has uncovered a fantastic regime of paradoxical, counter-intuitive quantum properties which do not fit into the classical scientific world view. The notion that physics was 'determinate' proves to be a reductionist fantasy. But careful contrivance is usually required to make quantum effects directly apparent and observable. Generally speaking, they just add another term to our incorrigible computational ignorance. Indeed, although they defy direct causal explanation, they are perfectly predictable statistically. In spite of popular belief to the contrary (and with due deference to their disturbing philosophical paradoxes), quantum theory and relativity are not somehow 'irrational' and could still be easily accommodated within the 'modern' scientific world view.

Recent research, however, has revealed a more profoundly indeterminate feature of the world.[7] Statistical techniques only work for systems that are microscopically *chaotic*. But a 'non-linear', or 'complex' system—and that includes almost every natural system when we consider it carefully—sometimes has the peculiar property of falling into a stable, macroscopically ordered regime. From a statistical point of view, such regimes are very improbable. Nevertheless, under certain conditions they appear spontaneously, without external prompting,[8] and seemingly able to reorganise their own order.

The appearance of spontaneous order in such complex systems is a widespread natural phenomenon. It is exempli-

[7]   Auyang, S. Y., *Foundations of Complex-System Theories: in Economics, Evolutionary Biology and Statistical Physics*, Cambridge: Cambridge University Press, 1998.

[8]   e.g. Cohen, J., & Stewart, I. *The Collapse of Chaos: Discovering Simplicity in a Complex World*, Harmondsworth: Penguin, 1994.

fied in turbulent flow, biological rhythms, mineral textures, geomagnetic cycles, etc. — perhaps even in the original emergence of life itself out of the primordial 'soup'.[9] What is so striking about such structures is that their form and incidence is genuinely unpredictable. Given nearly chaotic conditions, they can appear at any point, at any moment, as if out of nowhere. Everybody nowadays knows the fable of the hurricane nucleated by the beating of a butterfly's wing. An apparently normal event can thus have wildly improbable consequences.

The most complex of natural entities are, of course, living organisms. Biological species are also notoriously diversified. This is because they only approximately reproduce themselves. This opens them to biological evolution. 'Darwinian' processes of variation and natural selection adapt successive generations more and more closely to more and more diverse ecological niches. But they also permit the emergence of quite novel traits. One out of a billion bacteria just happens to have mutated to a form that is resistant to an antibiotic: it survives, and reproduces, and within a day it may have reconstituted the whole population.

The realm of the unpredictable turns out to be much more extensive than we had ever realised. The science fiction story tells of a time traveller, carried back to the Jurassic, who breaks one of the fundamental regulations of the Corporation and kills the proverbial butterfly. When he returns to the twentieth century, New York is not there any more. The moral is not just that New York came to Manhattan Island by an extended sequence of unscheduled events that might have turned out quite differently. It is that the very concept of a city such as New York is, to some extent, a product of historical chance. That goes for almost everything in our environment, natural or social.

The study of complexity and diversity is thus introducing people to quite new modes of scientific thought. The tradi-

[9]   Kauffman, S.A., *The Origins of Order: Self-Organization and Selection in Evolution*, Oxford: Oxford University Press, 1993; *At Home in the Universe: The Search for Laws of Complexity*, London: Viking Press, 1995.

tional terms of scientific discourse — laws, causes, properties, functions, classification schemes, etc. — no longer apply. They imply more clarity and logical necessity than is justified in practice. Actual natural and social systems are just too complex. Some of their characteristic modes of action defy formal prediction. They can be more or less explained rationally or roughly simulated computationally — but only retrospectively. That is, their outcomes are genuinely 'indeterminate'. There is no way of knowing just what will happen until it has actually taken place, as a real event in real time.

The world view that we have inherited from nineteenth century science was focused on a few limited areas of knowledge. Nevertheless, it was fondly supposed to apply in principle to the whole of Creation. *Unification* was its grand agenda. The 'project' of modern science, from the Enlightenment onwards, was to overcome 'irrationality' by the unlimited extension of orderly knowledge. But now this project seems to have proved itself logically impossible to complete.

It is not just that every answer spawns new questions.[10] Wise people have always recognised that perceived ignorance grows more rapidly than the knowledge gained in the effort to control it. What we now realise is that no amount of research can overcome chaotic or evolutionary unpredictability. Having demolished determinism and rubbished reductionism, we lack firm ground on which to construct an overall model of the Cosmos and all that is therein.

---

[10] Kant, I. *Prolegomena to any Future Metaphysics*. This reference is from: Horgan, J. *The End of Science: Facing the limits of knowledge in the twilight of the scientific age*, New York, NY: Addison Wesley, 1996., who got it from Rescher, N., *Scientific Progress*, Pittsburgh PA: University of Pittsburg Press 1978.

The post-modern critique

So scientists are at last beginning to realise that the idea of a 'Theory of Everything' is indeed just a dream.[11] For students of politics, philosophy, sociology, literature, culture, etc. this is an unmistakable symptom of the end of 'modernism'. In their jargon, there is no scientific 'Grand Narrative'. Science is as open to critical 'deconstruction' by the postmodernists as is any other body of human knowledge.

This scepticism is not unjustified. As we have seen, scientific knowledge is never unchallengeable in principle and often has to be revised in practice. The 'post-modern' critique is a healthy counter to philosophical positivism and political technocracy.

Unfortunately, post-modernism is frequently pressed much further. The fading of the beatific vision of a nirvana of Absolute Truth is taken to imply that all knowledge claims are equally credible, or equally dubious, or acceptable only to the members of a particular culture. This whole argument is clearly very schematic. Nevertheless, people who resent science in general, or dispute its findings in particular, fasten on it and deploy it as a formal justification of their own position. The role of science in society is thus significantly affected by the widespread dissemination of this 'lite' version of the post-modern philosophy.

But of course this reasoning is not logically compelling. In fact, it is indeed an intellectual cop out. It simply lumps together all the actual considerations that go into what we believe, or ought to believe, and dumps them out of the window. To some extent it ignores observed facts, experimental results, rational analysis, theoretical inferences and all the shared knowledge we gain from practical experience.

As I have shown at length elsewhere,[12] the particular virtue of science is that it provides a social mechanism for the systematic collection and comparison of the grounds for

[11] Barrow, J. D. (1991). *Theories of Everything: The quest for ultimate explanation*. Oxford, Oxford UP.
[12] Ziman, J. M. (2000). *Real Science: What it is and what it means*. Cambridge, Cambridge UP.

belief in specific propositions. In some situations, at least, these can be tested against one another and the inconsistencies resolved. Representations and narratives that survive this process surely merit somewhat higher credibility than those that have not thus been screened. Of course if hard-line post-modernists don't want to know about such matters, they are welcome to live their lives as best they may without them. But real people in the real world often find much benefit from scientific knowledge, where it is available to them, and need a serious account of just how much of it they can safely, usefully and excitingly believe.

The important message of post-modernism, then, is that scientists should always be questioned about the reliability of the knowledge they produce. This questioning cannot be brushed aside with the universal dictum that 'Science is the Truth'. Nor can it be trumped by a vacuous truism such as 'Science is just a Social Construct'. It calls for more detailed, more measured answers, covering specific areas of the world map. To put it bluntly: the 'role of science in society' depends very much on which bit of science one is talking about, as well as the social scene where this part is to be performed.

The globalization of knowledge

So what is the current state of scientific knowledge and its goals? At this point, I must admit, I am entering relatively unexplored intellectual territory. The trouble is that for most scientists it is almost sacrilegious to question the ultimate aim of a completely unified theory. So that there are no generally agreed alternative answers to public questions about the actual goals and achievements of the whole enterprise or of its major components. For a while it was said that the demise of the Superconducting Super Collider—by the 1993 decision of the US Congress not to proceed further with an extraordinarily expensive particle accelerator—signalled the End of Science.[13] This was clearly

---

[13]   Horgan, J., *loc. cit.* 1996.

just as foolish as the notion that the fall of the Berlin Wall signalled the End of History.

Nevertheless, it was an indication to scientists themselves that basic research is entering a new epoch. Although its implications for technoscience are of ever-increasing social significance, basic research is no longer activated by the 'universality' principle of 'modernity'. The acquisition of orderly knowledge is no longer justifiable as a many-pronged advance towards a single goal. We can no longer quite believe that the scientific project will eventually bring all of existence within its scope. So it is no longer plausible to chart our progress on a hypothetical map leading to that literally inconceivable Grail.

My impression is, however, that although modern science is not 'universal' it is genuinely 'global'. That means that it is no longer instructive to talk about our knowledge of the natural world as if it were, so to speak, an archipelago, or even an island continent, surrounded by an ocean of ignorance. Instead the Great Chain of Being turns out to be a loosely connected network encircling the whole knowledge sphere.

When scientists try to define their ignorance on almost any subject, they usually find that it is only partial. Instead of just hinting at enigmas, they discuss anomalies and theoretical contradictions, or they suggest that apparent blockages might be circumvented by approaching the problem from another direction. In effect, the domain of potential scientific knowledge no longer stretches into unbounded darkness. Like Planet Earth, it is finite to an overall extent, even if it does not have an 'edge'.

We should never forget Donald Rumsfeld's dictum about knowledge about terrorism: it is what we don't know that we don't know, which is worrying. But it may take another round of biological and social evolution before those completely unsuspected truths begin to emerge on this planet.

In the mean time, systematic research is producing quite detailed maps of what seem to be the most significant areas of the 'globe' of reliable, codifiable knowledge. The larger blank spaces, such as the nature of consciousness or the well-springs of social behaviour, are still poorly mapped. But this is because they are difficult to penetrate, not because we lack the wit or the will to explore them. From aerodynamics to zymology, scientists have methodically cleared the forests covering their various patches of knowledge and proceeded to cultivate an understanding of their special subject matter.

What is more, the cultivated fields of knowledge do not combine with one another neatly into self-contained regions. Many of the boundaries between the cleared patches are quite open, and the whole domain is criss-crossed with the webs of theory and technique. The traditional concept of an academic 'discipline' as a clearly defined 'nation-state' of knowledge, is losing force. Theories and methods overfly their frontiers. 'Transdisciplinary' paradigms produce and transfer knowledge freely between them. A great variety of experimental methods, research instruments, mathematical and computational algorithms, classification schemes, and explanatory metaphors find applications in apparently quite distinct scientific fields.

As we have seen, the conventional academic disciplines, like most nation-states, are somewhat arbitrary social institutions. At best they correspond, very loosely, to natural 'features' of the world. Historically speaking, they arose out of the need to create and maintain domains of order and rationality amidst surrounding regions of wilderness and intellectual anarchy. Like nation-states and capitalist enterprises, they can only survive by struggling continuously to grow.

In a global world, these expansive aspirations can no longer be satisfied constructively. They make no sense for the planet as a whole. So although the traditional disciplines remain convenient teaching modules, their boundaries are no longer strongly contested. Indeed, they are generally regarded as regions of ignorance, requiring to be opened and developed rather than defended. This applies particularly to so-called 'Mode 2' research, whose problems arise in 'contexts of application'. Scientists from different disciplines cannot properly perform their social role nowadays unless they actively cooperate rather than tacitly compete.

Scientific pluralism

Nevertheless, the 'maps' produced by transdisciplinary research are never really global in scope. Our scientific knowledge is wider than 'local' but seldom more than 'regional' in extent. A *plurality* of different paradigms, theories and classification schemes is required to express all that is known to science. This is because each such 'map' can represent only the particular entities and structures that happen to have emerged in a relatively limited domain of being.

Is this geographical metaphor getting a bit far-fetched? Let me try to explain in a bit more detail what I am getting at.[14] To start with, I will set aside the post-modern critique. In practice, nobody really doubts that science provides society with a great deal of very reliable knowledge. This includes robust representations of numerous naturally occurring or socially prevalent entities. It is futile to dispute the reality of galaxies, stars, planets, oceans, continents, organisms, neurosystems, ecosystems, artefacts, cultures and nation-states. It is also difficult to exclude less tangible entities such as atoms, electrons, molecules, genes, proteins, alphabets, money, and so on. Vast bodies of well attested knowledge have been accumulated about their properties and behaviour. In effect, they are the focal objects of whole sciences or scientific disciplines.

What scientists also tell us is that these entities are not just 'social constructs'. They are not contrived, or artificial. On the contrary, they have emerged 'naturally' in the course of deep time. Living organisms, for example, simply did not exist when the Earth assembled itself from a rocky halo around the Sun some 6 billion years ago. Just how 'life as we know it' appeared on the scene is still a very open question. But this does not seem to have been an event that could have been predicted from even the most meticulous knowledge

---

[14]   Ziman, J. (2003). 'Emerging out of nature into history: the plurality of the sciences.' *Phil. Trans. R. Soc. Lond.* A 361: 1617-33.

of the state of things before it actually occurred. Or even if it could have been foreseen in principle, its actual course and the precise nature of its outcome were not pre-determined.

Notice that I am not saying that living organisms do not have to satisfy all the laws of physics and chemistry. I am saying that they have properties — 'holistic' properties if you like — which are not predictable, or even expressible in terms of those laws. For example, the property of being a 'parasite' is well understood in biology. Yet this property is simply impossible to define, explain, or preconceive in the terminology of chemistry or biology.

The 'Life Sciences' are devoted to the study of these naturally occurring entities. To represent them systematically and intelligible, they had to develop an appropriate 'language'. This language is now wonderfully elaborated. It enables scientists to express clearly the immense range of phenomena, theories, research techniques, and practical results that they have discovered about the living world. But many of these phenomena etc. are simply not to be found at all in a purely physico-chemical, or biological worlds. So this language is not already contained in, nor 'reducible' to, the languages of the 'Physical Sciences'. Of course they have numerous words in common, but they are certainly not just local dialects of a universal and perfectly rational tongue.

The general public is thus not mistaken in sensing that there is a genuine gap of understanding between these two branches of the 'Natural Sciences'. They will not have it, for example, that plants can be 'engineered', or that animals are 'just machines'. They accept that the atmosphere is 'chemical' and can be changed to suit our convenience, but they regard natural eco-systems as unique and sacred. And they are not easily fooled into thinking that mobile telephone engineers really know whether their microwave transmitters might have some effect on living cells.

Of course many scientists are challenged to try to bridge this gap. That is the fascination of their ingenious computer models of 'Artificial Life'. Perhaps, in the end, their efforts will succeed. In the meantime, however, the social role of science is seriously compromised if it is presented as a unitary body of knowledge. On the contrary, its pluralism is a wise reminder that the world has many different aspects and cannot possibly be comprehended from just one point of view.

The reflexivity of the human sciences

These considerations apply with particular force to the 'Human Sciences'. For a variety of reasons,[15] they must surely be included in the 'global' archive of reliable, codified knowledge. For example, human beings are biological organisms as well as social actors. It seems unreasonable to differentiate in principle between systematic research which treats them solely as the former, and systematic research which treats them as the latter.

And yet, these two different approaches, although equally 'scientific', often yield quite different results. Indeed, these are frequently so disparate that they can only be expressed in completely different terms. Many of the phenomena, theories, research techniques and practical results obtained in the human sciences cannot be simply expressed in narrowly 'biological' categories. Just ask yourself whether the concept of a 'contract', say, has any real equivalent in a non-human context, and you will see what I mean.

So the plurality of the sciences shows up as another major knowledge gap. This gap, also, is a product of the prehistoric past.[16] It is associated with the evolutionary emergence of human *consciousness*, more than a million years ago. Completely novel entities and processes, such as social institutions and the symbolic media of communication, came into being at about this time. Don't ask here how this happened. Just note that it was an entirely unprecedented event, at least within the region of the Solar System.

---

[15]   Ziman, J. M. (2000). *Real Science: What it is and what it means.* Cambridge, Cambridge UP.

[16]   Ziman, J. (2003). 'Emerging out of nature into history: the plurality of the sciences.' *Phil. Trans. R. Soc. Lond. A* 361: 1617-33.

We are so much products of the social world that we take it for granted. But in making it a subject for scientific study, scholars have had to develop a systematic terminology. How otherwise could they discuss, compare and codify their findings. The 'language' of the human sciences is not, in fact, as well standardised, or as distinct from everyday speech, as its biological and physico-chemical counterparts. Nevertheless, it cannot be 'reduced' to, or translated word for word, into either of these other tongues.

Once again, the general public is well aware of this distinction. Most people make quite competent applied psychologists and sociologists. Indeed, they might often be wise to have more confidence in their ordinary or extraordinary 'common sense'. It is surprising how much attention they give to earnest academics who tell them that their ideas and desires are illusory concomitants of pre-conditioned reflexes, or that their behaviour is completely determined by their genes. Ironically, this sort of talk often comes from just the same scientists who urge people not to be taken in by astrology or 'creationism'.

Scientists, too, need to be cautious about how they walk across this gap. Indeed, some natural scientists so distrust the social sciences that they exclude them from their academies. Or they only include economics because it claims (incredibly!) to be a 'hard science, just like physics'. Such puritanical scientism misses the point. Scientific pluralism acknowledges the validity, propriety and credibility of many different ways of producing reliable knowledge. There are bound to be difficulties even with quite modest transdisciplinary exercises, such as applying 'biological' models — e.g. Darwinian evolution — to 'sociological' phenomena — e.g. technological innovation.[17]

It has to be said, however, that the human sciences are different in one very important respect. They are necessarily *reflexive.* They are engaged in the study of the very dramas

[17] Ziman, J. M., Ed. (2000). *Technological Innovation as an Evolutionary Process.* Cambridge, Cambridge UP, Wheeler, M., J. Ziman, et al., Eds. (2002). *The Evolution of Cultural Entities.* London/Oxford, British Academy/Oxford University Press.

in which they are themselves playing significant parts. Their 'findings' are amongst the 'makings' of the circumstances they claim to be representing. This reflexivity is unavoidable. Human actions are meaningful to those who undertake them. They can only be fully understood by the actors, and that is 'subjectively'. But this self-consciousness inevitably stands in the way of their sincere desire to be 'objective', at least to some extent.

The place of this whole domain of knowledge production in society is thus peculiarly complex. Take again the case of economics. Few professional economists, even in academia, are aloof theorists or dispassionate data collectors. Some of the best of them — John Maynard Keynes and John Kenneth Galbraith are prime examples — were, for periods in their lives, responsible government officials. The private advice and public utterances of professors of economics have considerable influence on economic affairs. So even when presented as 'non-instrumental', their scientific findings cannot perform this role in its ideal, disinterested form. Account has to be taken not only of the diversity of social values and personal goals by which these findings might have been shaped, but also of their effect on these values and goals.

In effect, the pluralism of modern society induces a corresponding degree of pluralism in its knowledge of itself. Neo-classical economic theory may be all right for industrial capitalists, but lacks credibility for subsistence farmers. The globalism of the world financial system is imposed on a patchwork of family, local, national and regional webs of production and exchange. Each of these operates according to its own rules, which have to be captured by scholars who are alive to and respect them.

That is probably the reason why so many sociologists and anthropologists insist that science is 'culturally relative'. Indeed, this is often true of the meaningful entities in their own disciplines. To take a naïve example, the concept of 'caste' in India, does not translate into, say 'social class' in England, or 'mana' in Polynesia. And as we move away from what seem to be the universals of human existence, the knowledge we produce is bound to become less and less general in scope. The human scientists are thus tempted to extend this characteristic of their own work over all the other sciences, forgetting how different these are in their subject matter.

Science: the infinite inner frontier

From the point of view of society at large it is essential to recognise that the human sciences do differ significantly from the other sciences. As we shall see in the next chapter, this has a major effect on the social role that we should assign to them. But the same applies to the distinction between the life sciences and the physical sciences. People should not take the 'globality' of present-day scientific knowledge as a sign that it is joined-up neatly everywhere. No amount of ingenuity can map over the natural rifts in its topography.

But this revision of the agenda of 'modernity' does not mean that the world has entered a state of intellectual decadence, where anything goes and nothing works. The search for understanding and control is not about to end in a bang of anti-scientific religious fundamentalism or in a whimper of post-modern nihilism.

Anyone familiar with contemporary scientific research will agree that it is as rigorous, vigorous, demanding and rewarding as ever. Its rationale is firmly rooted in the realities of an overpopulated, underdeveloped, but magnificent world. Research energises innumerable different kinds of knowledge domains, and is associated with many other intellectual enterprises. Although individual projects are sometimes organised on a global scale, they are undertaken by a great variety of independent enterprises and institutions. These have very diverse goals, ranging from the production of highly 'academic' knowledge 'for its own sake' to the most utilitarian aspects of technoscience.

Post-academic science is thus a fundamentally pluralistic social activity. For that reason, there is little practical incentive to unify formally all the knowledge it produces. Thus, even the traditional academic disciplines are not being incorporated into one big super-discipline. Metaphorically speaking, the established nation-states of knowledge are not being merged into a world-wide super-state.

But scientific knowledge is no longer a patchwork of independent sovereign disciplines. It is evolving into a comprehensive network of nodal domains. Each of these is a more or less specialised area of knowledge, representing, perhaps, a distinctive aspect of nature. But like the globalised world economy,[18] these are linked up every which way. They are connected, one might say, by the trade routes of methodology, the financial webs of theory, the population flows of expertise, the transport links of instrumentation, the oil pipelines of data and the electric power grids of technique. Amongst them, more 'local' problem areas flourish, like industrial regions of the mind, importing research technologies and exporting specialised knowledge for the whole 'globe'. But there are also, as we have seen, substantial areas of relative ignorance and major gulfs that may never be bridged. And despite all our philosophical wishful thinking, there is no law-giving authority governing all the fields of science or technoscience.

The established scientific disciplines have not actually withered away. They still play an active role in education, professional scientific careers, archival classification schemes, etc. They also continue to function as intellectual frameworks for non-instrumental research. But their conceptual schemes are traversed by a matrix of other knowledge structures. Thus, for example, the clouds of scientific questions surrounding climate change are notably interdisciplinary. They cannot be 'reduced' to puzzle-solving exercises fitting neatly into conventional academic pigeonholes.

What the public often fails to realise is that scientific knowledge is nowhere near as 'systematic' as scientists often pretend. It has many apparently arbitrary features. As we have seen, these are often the irreducible vestiges of past

[18]   Albrow, M. *The Global Age: State and Society beyond Modernity,* Cambridge: Polity Press, 1996.

historical events — cultural, biological, terrestrial and even cosmic. A formal 'top-down' perspective could never do justice to its amazing diversity. Nature, like 'human nature' looks different from different points of view on the ground. Its significant features cannot be wished away. But our representations of them are 'relative' to the standpoint of the local residents – that is, of the scientists who have made a special study of that particular aspect of reality.

The social role of the concept of a 'gene', for example, depends very much on how it is depicted. Although this is usually presented to the public as a unitary natural entity, ecologists, molecular biologists, evolutionary psychologists and pathologists each envisage it differently.[19] Can this plurality of versions be condensed into a single image for popular consumption? Our scientific understanding is enriched by the connections, comparisons, translations and analogies that we can make if we are able to travel around from one locality to another. But if there is a 'whole picture' — a single thought in the Mind of God — our intellects will be far too weak to take it in.

The transition to 'epistemic globalism' is thus a real break with the past. It already threatens the traditional disciplinary allegiances of most working scientists. Its unsystematic indeterminacy puts paid to the Enlightenment dream of a 'world government' for the sphere of scientific knowledge. At the same time, the thought that this sphere has now been charted in broad outline cannot challenges the 'endlessness' of the research frontier.[20]

This last concern, however, is exaggerated. As so often in the past, it may be entirely mistaken. But even if it turns out to be sound, it does not imply the end of the scientific enterprise. Consider, again, the political metaphor. Governmental globalism has no place for an 'external' frontier through which nation-states strive to expand. Paradoxically, that

[19] Rose, S. (1997). *Lifelines: Biology, Freedom, Determinism*. London, Penguin.

[20] Bush, V. (1945). *Science — The Endless Frontier: A Report to the President on a Program for Postwar Scientific Research*. Washington DC, National Science Foundation.

permits the emergence of a multiplicity of more permeable 'internal' frontiers surrounding innumerable localities still requiring socio-economic development. By analogy, the same applies to 'global' scientific knowledge. It may be, as many thoughtful scientists now believe,[21] that we already have a good grasp of most of the 'over-arching paradigms' of the major traditional disciplines—atomicity, quantum theory, evolution, DNA, and so on. Everywhere, however, within and between these clearings in the forests of ignorance, there undoubtedly are wonderful discoveries still to be made.

As we have seen, this 'localised' knowledge is not always purely utilitarian. It too has vital non-instrumental roles. Moreover, the geographical metaphor does not do justice to its richness. In a topological sense, each little 'area of knowledge' is a 'fractal' whose other 'dimensions' extend into infinite depths of complexity and diversity. Until these have been properly explored, if that is possible, we can have only vague inklings of what they might reveal. Much of the vast mass of knowledge to be thus gained will surely be abstract and esoteric. But a great deal of it will be inseparable from its social context. In other words, the frontiers are opening up between 'science' and 'society', in the domain of knowledge itself as well as in the busy life-world.

---

[21] Horgan, J., *loc. cit.* 1996.

ELEVEN

Science In Post-Modern Society

The life-world and the social order

It is now time to talk about 'Society'. But what is that? What is the social domain where 'Science' always has a 'place', and what is the drama in which it always plays a 'role'? What is the organism for which it performs a 'function'? What is the machine of which it is a 'part'?

Well of course, society as such is an enigma. It is what we are totally immersed in, as fish are in water. It makes up, and is made up of, our 'life-world' — our living experiences as individual human beings.[1] We cannot look at it 'objectively', as if from the outside. The most we know about it is that many features of our own life-world are also shared with the conscious beings to whom these features relate.[2] So we can at least talk to one another about these 'social' characteristics that we observe in common. And as we have seen, knowledge thus acquired (or, if you prefer, 'constructed') is potentially 'scientific'.

Unfortunately, much of this potential is still latent. The social scientists are far from a consensus about the fundamentals of their subject. They are not yet in agreement about how social phenomena should be represented formally, or simulated computationally. There are many schools of

[1]   Schütz, A. and T. Luckmann (1974). *The Structures of the Life-World.* London, Heinemann.
[2]   Ziman, J. (2002). *No Man is an Island. Hermeneutic Philosophy of Science, Van Gogh's Eyes, and God: Essays in Honour of Patrick A. Heelan, S.J.* B. E. Babich. Dordrecht, Kluwer: 203-18.

thought, but none so compelling that they must necessarily be followed.

So I am not making any scientific claims for what I am saying about society in this book. I refer frequently, for example, to the 'social order'. But this is not a reference to some hypothetical 'model'. I don't have my own pet theory about how society works. I don't pretend to understand just how it is that people are induced to behave individually with such predictable regularity that their doings can be combined systematically to carry out collective actions. I just note that this pattern of behaviour is an everyday experience. Metaphorically speaking, it is one of the salient features of the social domain, a characteristic scenario in the social drama, a normal function of the body politic, a typical mechanism in the social machine.

What we have seen, moreover, is that scientific knowledge is deeply woven into the fabric of modern society. At every level, it is applied, invoked, disputed or just sought. Its technological products cradle and enhance our lives. It is the trump card in so many of our planning games and the clinching rationale in so many of our decisions. Does that mean that we have already achieved the 'scientised' society envisaged by the technocrats? Is science now so pervasive that it is just a particularly strong coloured thread in the whole cloth of social life?

This rhetorical cliché is misconceived. Quite a lot of the knowledge produced by science does indeed enter everyday speech. Scientific ideas are indeed explicit factors in the material and social practices of our culture. But as we have seen, they do not come into the life-world as originally formulated, and are further transformed by social use. Once upon a time, 'alcohol' was an esoteric scientific term and the distillation of 'spirituous liquors' a mysterious scientific experiment. Now they mean much else besides, and are deeply tainted with good and evil. It would dilute 'science' into a meaningless universal if we were to include in it every element of present-day society that it once engendered.

Indeed, the distinctive property of science is that it is a well-spring of reliable *new* knowledge. This it achieves primarily by concentrating its activities and codifying what they produce. This is not at all to belittle the vast amounts of tacit, unpublished or disconnected knowledge acquired and put to use in technological or social practice. But our interest in this book is with science as a well-defined *institution* with a recognized place in the social order. As we have seen, it is not nearly as internally coherent or as sharply defined as many non-scientists seem to think. But it is much more tightly organized and socially installed than many scientists would like us to believe.

Our primary concern, then, is how the flood of new knowledge produced by science is actually received by society. Who needs it? Why is it needed? Who gets it? How is it used? What influence does it have? Who controls its production? Who suffers from a lack of it? And so on. These are not questions about the nature of science. They are questions about the nature of the contemporary life-world, and of the social order that sustains it.

Complications and pluralities

What sort of society do we now live in? The twentieth century took pride in being *modern*. It strove to be rational in thought, word and deed. It was original and experimental. It was sceptical of traditional beliefs and ethically unfettered. It was on an up-elevator of *meritocratic progress*. There could be nothing problematic about the role of science in such a society. It was to be fostered and followed, propagated and promoted, for science was literally the 'role model' for the whole culture.

Needless to say, this was a technocratic fantasy. The dark thoughts and even darker deeds of that century are still on our conscience. It is not just that the ideal of modernity only had some substance for the richer people in a small part of the world, and was not meaningful for the great majority. As we have seen, its notions about the nature of scientific knowledge and its capabilities were misconceived. Enacted politically, they were bound to lead to social disasters. The Stalinists and Maoists called their revolutions 'experiments' — and murdered millions to get the 'correct' results!

We no longer accept the 'modern' account of the nature of scientific knowledge. Even amongst natural scientists, only a few now really believe that systematic research and analysis will reveal a unitary code of universal knowledge covering the whole of the natural world and the human condition. The version that is surviving the post-modern critique is in a much lower philosophical key.[3] Science is still generally believed to be a very effective means of producing extremely reliable knowledge, but its findings are no longer considered unchallengeable in coverage, coherence or validity.

At the same time, there has been a major transformation in the social organisation of knowledge production. New institutional forms, such as global technoscience and its post-academic side-kick, now hold sway. These clearly have an even more influential place in our culture than their predecessors. But they do not inspire enthusiasm as potential role-models for the good society. At best, they merely incorporate and exemplify certain general features of the present-day social order. Some authors seem to suggest[4] that society imitates science, just as 'life' sometimes imitates 'art'. But it is more plausible that the science of today mirrors its social setting, just as it is said to have done in the past.[5] Science is surely shaped by the general spirit of the age. Nevertheless, like other distinct cultural institutions, it also has its own agendas and imperatives.

Our culture as a whole has also changed markedly since the end of the Second World War. I refer here, of course, to the way of life of well-to-do people in the most economically advanced countries. But this favoured way of living is disseminated world-wide by television, mutates mon-

[3]   Langer, Susan (1957). *Philosophy in a New Key*. Cambridge MA, Harvard UP.

[4]   Nowotny, H., P. Scott, et al. (2001). *Re-Thinking Science: Knowledge and the Public in an Age of Uncertainty*. Cambridge, Polity Press.

[5]   Forman, P. (1971). 'Weimar culture, causality and quantum theory, 1918-23.' *Historical Studies in the Physical Sciences* 3: 1-116.

strously in vast urban slums, and dribbles down from local elites into the least developed rural areas. Its globality is not confined to the trade routes and industrial production chains controlled by multinational corporations or the planetary range of nuclear missiles. It might also be symbolised by an Inuit hunting seals from a snowmobile, a Tibetan monk on the internet through his cell phone, a Zambian woman too poor to buy an antiviral drug for AIDS, or a Mexican peasant leader quoting Gandhi and Mao. Such unlikely combinations of the old-fashioned and new-fangled, the meagre and extravagant, the material and spiritual elements, have their historical rationale. But the whole system can scarcely be described as 'enlightened'.

The fall of the Berlin Wall was the end of a political era. But it was not marked by an abrupt cultural discontinuity. Yet present-day society is very far from what was envisaged by the prophets of modernism, half a century ago. In a word, we now live in what can only be called a *post-modern* world. Quite apart from its possible philosophical resonances, what does that imply?

It is sometimes said that our culture is *post-industrial*. This suggests a decreasing role for traditional technoscience. But it underplays the immense scale of the material and human resources still devoted to the mass production and distribution of useful — and useless — artefacts. It might mean simply that it is a *consumer* culture, where now 'shopping' has become the predominant economic activity. That could explain the growing influence of the human sciences, with their supposed understanding of the arts of presentation and persuasion.

A more apt cliché is that it is an *information* society. The exchange of meaningful messages is now more significant than the trade in goods and services. True enough, the technology for the transfer, transformation and storage of data, texts, sounds and images has become stupendously more effective. As we have seen, these techniques have revolutionised the production and distribution of scientific knowledge. They profoundly affect the connections that sustain the social order. For all of us — including that iconic

Tibet monk — electronic devices are now normal features of the life-world.

It is questionable, however, whether the rise and rise of Information and Control Technology — ICT — is the defining characteristic of the post-modern world. Visual images in real time are indeed received differently from verbal messages or written texts. As Marshall McLuhan famously joked — 'the medium is the message'.[6] But instant, global communication via the electric telegraph, the telephone and radio broadcasting was already fully incorporated in the modern culture. Advanced ICT certainly facilitates the exchange of meaningful messages but does not directly determine what is communicated. Much of it is still personal gossip, crime reports and popular music, just as it has been for the best part of a century.

Say, rather, that ICT is just one of the products of technoscience that enables us to live much more *complicated* lives. I use this word, rather than 'complex', because I am not presenting society as a 'system' of which I have a 'model'. I mean, simply, that the life-world of most people nowadays has many more distinct and different elements than they had in former times. The daily round and the common task are more intricate and confused.

Everything we ordinarily do — getting and keeping a job, obtaining and preparing food, finding a home, getting married, having and raising children, taking holidays, saving for our old age , and so on, has become more of a palaver. It seems to require more careful thought, more significant choices, more attention to detail, than it did for our parents. So we need to seek a great deal more information to assist our personal decisions. This demand is insatiable, so it stimulates a wildly excessive supply. The 'information' surging around via ICT in our society could be cut by 90% simply by excluding junk mail, government handouts and mass media advertising.

---

[6]   McLuhan, M. (1962). *The Gutenberg Galaxy*. London, Routledge, Kegan Paul.

Less rhetorically, these life-world complications reflect the *pluralism* of the post-modern social order. It is often remarked that we now live in a *network* society.[7] In other words, we are each linked to many other individuals and institutions, and must adjust our actions to theirs. These adjustments require the continual exchange of information about each other's intentions. What do parents and teachers expect of one another? , what do shopkeepers and customers look for? How do plumbers and householders react to their stereotypes? Or what do insurance companies and policy holders think of each other? Do employers and employees have any mutual respect for each other? What if any is the rise of antagonism between priests and parishioners, between governments and tax-payers? or voters and politicians, or civil servants and citizens? What about the mutual feelings of universities and their alumni, doctors and their patients, charities and their donors? Looking back to the time when ICT was fresh and new it was thought that we were approaching paper-free offices But now, how much is still being required in the way of newsletters, account statements, handouts, telephone calls, public advertisements, emails, company reports, begging letters, questionnaires, user manuals, warranties, contracts, and timetables? The answer lies in the acres of small print that seem to be necessary to stabilise the network link between us and our outputs and inputs.

In past times, most of these relationships were governed by custom and convention. If there were problems, they could be dealt with verbally, face to face. In many cases, they were strictly hierarchical. Those at the top gave instructions, which those at the bottom had to obey. Modern industry was based upon the culture of the factory or the office, with masses of 'workers' ruled by a managerial bureaucracy. Reams of paper were often consumed in carrying information up the system but a few words sufficed for the decisions that were handed down. Only in the market place, amongst merchants and small proprietors, was it

---

[7]   Castells, M. (2000). *The Rise of the Network Society.* Oxford, Blackwell.

necessary for individuals to *negotiate* on equal terms with one another — that is, interact by exchanges of information until they had come to agreement.

As we shall see, post-modern society is not just one enormous market system. But it does claim to be subject to the rule of law. Individuals and institutions have prescribed rights and duties towards one another. Equity requires equality of standing, especially in access to relevant information. So all active social relationships are structured on that basis. It is not that conflicts are expected. But it is prudent to foster mutual understanding before they can arise. And the lawyers can be relied on to multiply the communications a hundredfold before they are completed.

In any case, most of our social transactions are with strangers. A large proportion of the links in the network are 'virtual'. Our health is covered, for example, by state or private insurance. We can vaguely envisage the circumstances when a connection with this node of the social network might have to be activated. But we didn't know, before we went into hospital, anything about the surgeon or the nurses to whom we now have to entrust our life. An immense amount of information has had to flow back and forth, along these and other channels, to ensure that this relationship could proceed satisfactorily.

The manner in which we communicate with one another also depends greatly on the context. With friends we are unaffected, with strangers polite, with officials formal, with clients professional, with customers complaisant, with competitors wary. Nothing is more disconcerting than a person playing the wrong social role — a son of the family behaving like a lodger, for example, or a professor like a student.[8] In Arcadia, a princess was born a princess, and performed naturally in that character for life. Similarly with a shepherdess. One of the major complications of post-modern society is the diversity of roles that each of us may be called on to assume. Our life-world is *multimodal*. We are

---

[8]   Garfinkel, H. (1967). *Studies in Ethnomethodology*. Englewood Cliffs NJ, Prentice-Hall.

continually having to switch from one version of a 'protean' self to another.[9] We need to keep our wits about us even to recall who we are.

From lore to literacy

I am not suggesting that elaborate and diverse media of communication were entirely absent before the modern era. Anthropologists report on the bewildering multiplicity of Hindu occupational castes, and on the intricacy of the kin-ship relations amongst the aboriginal Australians. Shake-speare —the actor poet—remarked on the 'many parts' played by 'one man in his time'. Molière's characters knew all about networking and dissembling. Dickens opened windows into a great variety of multiply-connected life-worlds where 'unconsidered trifles' of information were often of vital significance.

Nor am I suggesting that our personal lives are shaped any less inexorably by the human imperatives of 'hatch, match and dispatch'. Post-modern society is complicated beyond comprehension. Nevertheless its variegated life-worlds must accommodate to these absolutes. Most of us treat society as a 'peoplescape'[10] where we contrive to find or make comfortable enough niches for ourselves. Indeed, amongst its many 'modern' elaborations are the social welfare safety nets that we hope will catch us if we fall out of our nests.

What is quite clear, however, is that a lot of knowledge is now required to operate competently in this environment. One must be able to decipher the relevant communications, connect effectively with the appropriate nodes, perform adequately our allotted tasks, and so on. Much of this knowledge is tacit. It is acquired imperceptibly, by experi-ence. Social *praxis*, for example, doesn't make perfect. But having to work with other people is still the 'best' way to

[9]   Lifton, R. J. (1993). *The Protean Self: Human Resilience in an Age of Fragmentation*. Chicago IL, University of Chicago Press.
[10] Primavesi, A. (2000). *Sacred Gaia: Holistic theology and earth system science*. London, Routledge.

learn how to get on with them. Any sane adult who thinks they need a book on, say, how to win friends and influence people, is unlikely to profit from its advice.

In traditional societies, cultural practices were illuminated, embedded and enhanced by *lore*. Although typically uncodified, lore, the folk knowledge of the natural world, was often quite specific and reliable. As we have seen, it was the base for many later sciences and technologies. Getting to know it by heart and to use it with ease was a pre-requisite for personal survival.

The 'modern' way of life, however, is more demanding. Its messages, for example, are seldom delivered by word of mouth. They are typically conveyed in written form. A major part of its knowledge base is systematically collected, codified and stored in texts that are only intelligible to those who can read. Its technology is published in patents, its money supplies are enciphered into bank statements, and its legal decisions rely on written statutes and precedents.

So this is a life-world where basic literacy is an essential personal skill. What is more, it is a social system that could not operate if the great majority of the population could not communicate and calculate in symbols. Universal elementary *education* is at the top of the political agenda for any country seeking to modernise. The necessary knowledge has to be acquired in childhood by systematic *schooling*.

Nevertheless, it is surprising how very little additional formal *learning* an ordinary individual really needs to manage quite satisfactorily in the 'modern' life-world. Think of the factory workers and agricultural labourers, miners and dockers, shop assistants, clerks and servants who formed the great majority of the population at the beginning of the First World War. Think of them going off in their massed battalions to the mechanised slaughter of that War. Think of their equivalents now, in the urban sprawl of Cairo, or Calcutta or Sao Paulo or the new industrial zones of China or Mexico. The schooling of most of them was typically uninspiring and ended thankfully in their early teens. At best, they could read a tabloid newspaper and write a halting letter home. This skill did not involve abstract ideas.

But it was enough for them to pick up informally all the other things they needed to know. Apart from the popular press — and nowadays TV — this was still mostly learnt, as in traditional society, in the family circle or amongst their mates, on the job and in the street . But it enabled them to perform dutifully as productive workers, caring parents and loyal citizens.

How then could 'modern' society achieve such unprecedented technical sophistication? As we shall see, the answer is that its social order was dominated by a knowledgeable *élite*. Engineers, doctors, lawyers, civil servants, accountants, academics, clergy and other 'professionals' were normally university graduates, or were trained to exacting standards in the course of their employment. They were not necessarily the owners, or senior executives, or state officials who actually controlled the larger units of production. And they were supported by a 'lower middle class' of 'technical' workers who had received just sufficient 'further' education to perform with understanding their skillful practical tasks.

The power of all these elites rested on their knowledge bases. Between them, they had access to all of this that had been produced or was being produced, by science and technology. This knowledge was required to create and maintain the industrial corporations, the accountancy systems, the financial markets, the legal structures, the medical therapies, the state apparatuses, the church hierarchies, the military machines and other elaborate social institutions that spanned the modern world. But the ordinary people whose life-worlds they framed and sustained needed to know very little about how they really worked.

Adapting to change

Modernism is *technocratic*. It provides people with good lives provided that they conform to its technical necessities. The complications of their life-world knowledge can always be resolved by carefully following its written rules and practical instructions. They are not to be expected to understand the rationale of these rules or the techno-scientific basis of these practices. That esoteric knowledge is restricted to an elite of research specialists or licensed practitioners, whose higher education in the sciences has cost them much effort and many years of their lives.

This policy is only feasible in a static autocracy. As the last decades of the Soviet Union showed, it is also a recipe for economic stagnation. It ignores the intrinsic dynamism of technoscience. The paradox of modernism is that it worships 'progress' but has no means of coping with its effects. It educates people just sufficiently to deal with the complicated life-world of the day, but makes no allowance for the fact that the whole scene will have changed within a generation.

This, of course, is another cliché. Down the centuries, we oldies have always complained about the terrifying pace of social and technological change. In practice, though, healthy adult humans have always proved remarkably good at learning by and through imitation and experience. They adapt in body, mind and spirit to the new imperatives of their culture. My parents were born into the late Victorian world of the horse, bicycle, railway engine and steamship: they never seemed to have any problems with using the telephone, driving a motor car or travelling by air.

What is more, cultures can fight back. Revolutionary social innovations that push people too far from their customary ways often have to revert to more traditional forms. Commercial competition compels novel technologies to become 'user friendly'. Even the instructions for the remote control of a video recorder usually require no more than basic literacy — provided one can devote a whole afternoon to deciphering them! So the knowledge of our life-world hasn't become entirely unmanageable after all!

There is no doubt, however, that the pace has recently quickened. People are also living much longer. Thus, a whole dimension of rapid change is further complicating post-modern society. This applies in every sector of the life-world — health, food, drink, housing, employment, holidays, entertainment, 'consuming', insurance, banking, tax-paying, voting, and all, not to mention the basics of birth, mating and death. On the one hand, what we had assumed to be solid features of everyday life, which we had taken so much trouble to comprehend, eventually crumble inexplicably. On the other hand, artefacts and practices that had seemed trivial fancies suddenly become vastly important, and have to be mastered, pronto. Social barriers that we had come to consider immutable are dissolved, whilst others are raised where they had never been noticed before. Mergers are announced between institutions that had always seemed to be at daggers drawn, whilst old established firms and political parties unexpectedly rip themselves apart. The scenery around us is kaleidoscopic in motion, and not just labyrinthine.

As I say, this sort of thing has always been going on. But it used to happen in slow motion. The post-modern speed-up evidently has a number of causes. Growth through innovation is the driving force of technoscience. Globalised market capitalism sets a frenetic pace. The politics of an overcrowded, over-exploited planet are dire and hectic. To some observers, the whole system appears to be on the brink of dangerous chaos.

But I do not propose to speculate on why this has happened, or what could be done to avoid collective disaster. In any case, for the great majority of people in the world, this is neither bad news or good. They are just very poor and powerless, as they always have been. But even adequately resourced citizens of advanced countries are perplexed and frightened by the complexity and instability of their individual life-worlds.

To some extent, the prophets of modernism foresaw this development. Yes, they wanted to put science into the driving seat. Yes, this would lead to all sorts of problems. But they were so bemused by their technocratic visions that they thought that these could all be solved by just putting science under even greater control. This could be done, for example, by raising the technoscientific level of politics, economics and other moral disciplines. The social sciences would soon get their Newton, and all would be light.

That was what Charles Snow really meant when he lambasted the humanists for being so ignorant and scornful of the natural sciences.[11] How could mandarins educated solely in the classics or history be expected to cope with technical and social innovations whose basic principles were like magic to them? Were they likely to generate, install, or even favour, such untried novelties? Well, if they didn't understand these things, they should shift over and hand the reins to the 'new men' who really did.

More soberly, Snow was for giving science much more leverage in society. Like his contemporaries, he valued it for the health, wealth and happiness it would bring to society. Since his day, that has become the official ideology of all advanced nations. The only change is that nowadays the stress is all on the wealth that technoscience creates. Even health is metered out economically. The 'non-instrumental' functions of scientific knowledge are simply ignored.

Nevertheless, this general line of argument has certainly helped to promote a broader educational curriculum for the whole 'professional' elite. It was also used to justify much wider educational reforms. An essential feature of the 'information society' is mass secondary and tertiary education. Compulsory schooling now extends to the late teens and higher education has expanded to include something like half the relevant age group. How otherwise could individual citizens be expected to receive, act upon, transform or re-transmit, the complicated messages that flow through their nodes in the social network?

What is more, *technical* education brings active scientific knowledge to a great many more people than in the past. A significant proportion of the population — of the order of 5%, I would guess — have everyday jobs that use relatively recent scientific ideas. Engineers, architects, doctors, dentists, veterinarians, lawyers, school teachers and other professional practitioners have long been required by law to pass university examinations on large chunks of the basic

---

[11]　Snow, C. P. (1964). *The Two Cultures: and a second look*. Cambridge, Cambridge University Press.

sciences of their craft. But similar formal educational qualifications are now obligatory for much larger groups of skilled workers.

Needless to say, this applies to the great majority of the people working in technoscientific enterprises. The members of a research or design team don't necessarily have PhDs, but they would almost all be counted officially as QSEs — Qualified Scientists and Engineers. Formal educational qualifications are now required in many other callings. The training of a hospital nurse or a school secretary for example, include biomedical theory as well as systematic practice of filing information on the all-pervading computer. Social workers are expected to have a smattering of academic psychology and sociology. Even the most mundane curricula in business studies present economics and informatics as if they were scientific disciplines.

Few of the people who receive this sort of education would describe themselves as 'scientists'. They get their science at second-hand. Most of them are not directly employed in the knowledge production industry, and only have brief, tangential contacts with it. As we have seen, it is highly desirable for them to be taught by people who are actively engaged in non-instrumental research, but this is only feasible at the university level. They don't pretend to achieve a high standard of scientific understanding, and mostly they do not try to retain or extend what they have learnt in later life. But they form an 'attentive public' to science and technology and watch its progress with interest — and not just ill-informed concern.

In sum, the post-modern 'person in the street' has had much more formal education in the basic sciences than any previous generation. They are thus much better equipped to cope with a life-world whose complications are so largely scientific. People nowadays can do more than just decipher written information and instructions. They can appreciate the significance of this information and carry out instructions more knowledgeably — that is, more intelligently and with more social understanding. So, instead of just reacting to events as best they can, they are enabled to grasp the meaning of foreseeable developments, and to some extent they can prepare themselves for their practical consequences. They can thus begin to shape their lives more actively, tackling problems in principle before they have to be untangled in real time. In a word, through possession of this knowledge they can become less the victims of circumstances, more the masters of their own fate.

Getting to 'understand' science

And yet the expectations of the technoscientific elite have been disappointed. The general expansion of secondary and tertiary education has not overwhelmingly favoured the natural sciences and their associated technologies. Although the total number of students specialising in them has greatly increased, these disciplines have not grown disproportionately. They do not dominate the academic league table. The very cleverest pupils still compete ferociously to demonstrate of their precocious mathematical talents. But then they go off to make amazing fortunes in the City. Why are they not slaving away at a pittance making amazing discoveries or returning dutifully to the schoolroom to teach the next generation?

The enthusiasm of leading scientists for ' the public understanding of science' articulates and operationalises this disappointment.[12] Their very patronising term for this is 'scientific literacy', with its implication that what is missing is culturally indispensable, and that they, the scientists, are the *literati* who can make up the deficiency. For them the place of science in society has not changed. They have always had the ball in their court, and they are not going to give up its possession now.

Thus, from the point of view of the post-modern citizen, the place of science has not greatly changed from what it was a century earlier. It is still officially fostered and promoted rather than popularly boosted and celebrated. It is still the product of a cluster of distinctive, and often rather

---

[12]   Bodmer, W. (1985). *The Public Understanding of Science.* London, The Royal Society.

distant institutions. It still reaches the life-world with its tone of authority. It is still owned by 'them', rather than by 'us'.

The emergence of a technical stratum of people with various degrees of scientific training has not greatly changed the situation. It certainly does not prove that science is fully incorporated into the post-modern culture. Political and economic forces herd the 'scientists, technicians and managers' together into a big trade union, but they are not really a coherent, self-conscious social *class*. In career terms, they count themselves as members of organised *professions*, not of scientific *disciplines*. They defer to the academic and technoscientific elites, but their primary allegiance is to the needs of their clients, patients, customers or employers, not to the imperatives of 'scientific truth'. If it ever came to a show-down, they would say that they belonged with 'society' rather than with 'science'.

Nevertheless, this technical stratum does function as a bridge, or buffer zone, between high science and the public in general. Even if they are not positively enthusiastic for science, its members do not dismiss it as incomprehensible, or irrelevant, or 'boring'. Their work often requires them to act as mediators, or interpreters between various scientific and life-world domains. And they are also visible as ordinary fellow-citizens, family friends and relations, parents of our children's school friends, fans of the town football team, and so on. Universal science education enables the rest of us to test their human normality against popular culture, and to share with them many aspects of the life-world.

In these and other ways, science has ceased to be entirely esoteric. Nor is it pure sorcery. Everybody has had to actually *learn* a tiny bit of it at school. It gets quite a lot of attention in the newspapers, soberly as well as sensationally. It is picked up from the radio and telly in news items. Feature programmes and glossy magazine articles, reveal scientists and their habitats to curious eyes. A few of the many excellent popular books about it become best-sellers. Science and technology museums are visited by millions of children and figure on the regular lists of tourist attractions from knowledge about the village mill, to the back-street slums of Victorian times and how people gradually learnt to use the new inventions of the day.

By formal academic standards, of course, the outcome of all the efforts to make the general public 'understand science' is pathetic. People still can't give correct answers to the most elementary questions. Does the Sun go round the Earth? or the Earth around the Sun? But these deficiencies mainly shows how clumsily the teaching of science has been incorporated into our educational system. It is still too subservient to the special needs of the professional elite and their technical ancillaries. However much people enjoy trivial quizzes, they resent being humiliated by examination questions. The secondary school science curriculum is still differentiated into the traditional 'disciplines' — 'physics', 'chemistry' and 'biology' — and presented as if every pupil were competing for a research career in some specialised corner of one of them.[13] Naturally enough, 99% of them eventually fail at that level, and make better use of their talents elsewhere. But having been defeated by school science, they retain little recollection of, or affection for, its actual substance.

Indeed, people often leave school asserting that scientific knowledge is just a load of 'facts'. Scientists find this very puzzling. The beauty of science is that its facts are intricately structured by theories. In the end, we have learnt to look with X-ray eyes through the flesh of facts to the amazing skeleton within. The trouble is, though, that these theories are built up layer by layer. Although each new layer is based on the ones below, it tends to be presented initially as if it were a new body of 'facts'. So until the whole theoretical framework is shaken into logical coherence — at the climax of many years of learning — it does seem to be overwhelmingly factual.

To put it another way: science does not have an elementary set of ideas, symbols and skills with which even its

---

[13] Ziman, J. M. (1980). *Teaching and Learning about Science and Society.* Cambridge, Cambridge UP.

most elevated products can be decoded, however imperfectly, by the near beginner. Getting to know it is less like achieving alphabetical literacy than learning to read Mandarin Chinese. Nevertheless, the ability to recognize a few hundred of the most common characters is presumably sufficient for many everyday purposes. As I have emphasized, although the post-modern life-world is pervaded by the knowledge of technoscientific products, new scientific ideas keep entering as from an alien culture. Reception of these ideas at a personal level is greatly facilitated by even a very modest vocabulary of frequently-used scientific concepts. It makes them seem friendly, and perhaps worth getting to know better.

The dissemination of elementary scientific ideas through universal schooling and media attention is thus a positive benefit. It is one of the distinctive features of an economically advanced society. Nevertheless, this achievement is of very limited scope. It does not, as some of the enthusiasts for the public understanding of science seem to suggest, induce people to favour the cause of 'scientific rationality' on every controversial issue. Nor, by contrast, does it 'empower the people' against tyrannical scientific expertise. It does not even dispel the religious superstitions, pseudoscientific doctrines, phoney therapies and other 'irrational' beliefs and practices against which established science has always stood so firm.

Risk in the post-modern life-world

The post-modern culture does not repudiate science. On the contrary, it fosters, respects, rewards, and responds to it quite as warmly as ever. But it does not, as the modernists hoped, incorporate it into its being. The mere fact that this is knowledge that has to be positively 'learnt', or 'understood' is a tacit reminder that it was produced elsewhere. It does not grow out of our own soil. It still comes into the life-world from another sphere, like the coffee from Brazil or the news of some dreadful earthquake in Iran. We don't have reason to disbelieve it, but we don't cling to it in moments of stress, as we do to the folk and street wisdom of the life-world that largely governs our lives.

Nowhere is this more evident than in the scientific treatment of *risk*. Some sociologists hold that we live now in a 'risk society'.[14] Perversely, scientific progress has the effect of making people more anxious about its dangers. It is not true that life nowadays is more hazardous than it was in the past. In fact, one of the genuine services of technoscience is that it makes the lifeworld safer than ever before. Well, we certainly live longer, if that is a relevant indication.

Nevertheless, the increasing pace of change faces people with novelties about which too little seems to be known. They have to make decisions whose outcomes they find it very difficult to assess. Nowadays, many of these uncertainties arise out of advances in science. Have the side effects of this new vaccine been fully explored? Will this new radio mast endanger my family's health? So we go back to science for more information.

Science also exposes hazards that we had not previously suspected. Is deadly radon seeping into our lovely old house? Shouldn't we pull out the Victorian plumbing because of the lead in the water? Wood fires are cheerful, but from their smoke we ingest noxious dioxins and carcinogenic carbonaceous nanoparticles. And what about the possibility of our earthly home being wiped out by an asteroid? You gave us the bad news, Professor: so now, please, let's have the good news—explaining how to jump free?

As we have seen, a bit of scientific knowledge is a great help in such circumstances. But that suggests that if we worked at it we would find that we could foresee the worst dangers and avoid them. This belief is not ill-founded. It encourages governments to make laws to prevent us from putting ourselves unwittingly at risk. It induces insurance companies to enforce safety codes for vehicles, houses, and consumer goods and the hand of God. It motivates trade unions to insist on rules outlawing unsafe working practices.

---

[14]　Beck, U. (1986 (1992)). *Risk Society: Towards a New Modernity*. London, Sage.

In effect, the unveiling of hidden risks easily persuades individuals to accept the collective *regulation* of their ways of living.[15] The highly regulated culture promoted by technocratic modernism is actually consolidated by post-modern insecurity. As we shall see, this puts an increasing burden on the state, both as the regulatory authority and as the source of knowledge identifying risks and justifying their control. It also requires continually improved public understanding of what is involved, and why they must do what the regulators command.

Unfortunately, risk perception and regulation feed on one another in a vicious spiral. Even what the most learned scientists say they know, turns out to be unreliable and incomplete. So the deeper we delve, the more risks we uncover, the more anxious we become, and the more we legislate to try to avoid them. Thus, the post-modern lifeworld which we have taken for granted for so long, becomes embedded in a disconcerting hybrid of a 'risk society' and a 'regulated world'.

And yet a 'risk' is essentially a scientific concept. It portrays a possible peril 'objectively'. It presents it as a communicable quantity, subject to a symbolic calculus. Even when we can't estimate it precisely, we can still be persuaded that it is greater or smaller than some other parallel hazard. In modern usage it derives from the social technology of insurance, but this is often combined mathematically with elements from a material technology such as engineering. Until the latter half of the twentieth century it figured mainly in technical discourse about the safety of aircraft or the actuarial value of life insurance policies. The notion that certain risks were 'acceptable' then was still not the commonplace that it is today.[16]

Does the percolation of the notion of risk into everyday talk indicate that the post-modern culture is really permeated with science? Not at all. What people perceive as a sig-

[15] Hagendijk, R. P. (2004). 'The Public Understanding of Science and Public Participation in Regulated Worlds.' *Minerva* XLII(1): 41-59.

[16] Ravetz, J. R. (1977). *The Acceptability of Risks*. London, Council for Science and Society.

nificant hazard is intrinsically subjective. It is a hypothetical event in their lifeworld, and thus inseparable from its social context. Its personal value, whether negligible or fearsome, combines a variety of local or historical factors that avoid calculation.

What is more, the notion of an 'acceptable' risk is largely a *cultural* construct.[17] Whatever the engineers, doctors and statisticians may say of a particular hazard, people behave according to socially established assessments of its dangers. For example, we are brought up not to flinch at the possibility of being killed in a road accident, even though it is customary now to be very anxious about the much lower 'objective' risk of dying in a train or plane crash. There are plausible explanations for this *non-sequitur*. But in the end, all the subtleties of the human sciences cannot do justice to the paradoxes and idiosyncracies surrounding this apparently simple term.

So even in this intimate lifeworld setting, where Science would seem to have the most to give to Society, its place is not entirely secure. The relationship is strained on both sides. They expect too much of each other. Society doesn't get all the *certainty* it was seeking: Science doesn't receive all the *trust* it feels it deserves.

Unfortunately, this disharmony is growing rather than lessening. Scientists and their technocratic allies complain, of course, that it is simply due to public ignorance. If people only understood the findings of science better, they would appreciate both their uncertainties and the compelling nature of those that are well established. Experience, however, does not bear this out. Futhermore, it seems that the scientific elite can only blame themselves for these deficiencies. Traditional science education, for example, completely failed to emphasize the inherent uncertainties of what little of it was taught. So whenever a 'scientific' risk assessment turns out to have been grossly in error — for example, over

---

[17]   Douglas, M. (1986). *Risk Acceptability According to the Social Sciences*. London, Routledge & Kegan Paul.

'mad cow disease' — people feel, quite rightly, that they have been short-changed.

In fact, despite official anxieties, the post-modern public is actually better instructed on the substance of science than any previous generation. The internet also provides unrivalled access to scientific knowledge, often in highly palatable, if not entirely reliable forms. Prominent political, legal and ethical issues have made people much more conscious of its limitations. But this healthy scepticism is not focused on the *technical* capabilities of the research process. It tends rather to take the form of *social* criticism of scientists and their institutions. In other words, it fuels the *distrust* of Science that now bedevils its relations with Society.

The trouble is that interested parties gather like vultures around 'risky' situations. Here is where academic science has always played a vital role. When disputes arise, it can usually show that it is not directly in the service of any of the contestants. People trust it as an *independent* source of knowledge and expertise. But as we have seen, the transition to post-academic science compromises that independence. People are given genuine reasons to suspect that it might not be as 'objective' as it claims. As a result, it is difficult to restore public confidence in scientific knowledge, however well-founded technically this may be. This is a non-instrumental social benefit that we simply cannot afford to lose.

A question that frequently arises when risks are perceived is whether the requisite scientific knowledge is available to deal with them. That is almost always a problem. In fact, it highlights the obvious defect in any policy purporting to teach people enough science at school for them to grasp and overcome the lifeworld hazards they will encounter. These risks, and the science that creates or reveals them, are changing too rapidly to be anticipated decades in advance. The wisest scientists of today are completely foxed by questions about what their own children will encounter when they are grown up. We all, scientists and the public alike, just have to continue our education, right through to old age. Learning is the signal of life, as concerns about Alzheimer's disease show only too well.

Even the best that scientists currently know may be quite inadequate. It may also be much too specialised. Technoscience has learnt to be cautious about the immediate 'side effects' of its innovations, but is not geared to consider their wider lifeworld consequences. Academic scientists have detailed knowledge about a great many aspects of things as they are, but seldom think about what might happen more generally if a particular change were made. All useful scientific debate about the risks of genetically modified crops, for example, has been on hold for some years until intensive research on agricultural ecosystems has produced reliable data on their actual effects. Now it seems the common potato has, maybe, won the case. Societal mechanisms are clearly needed to initiate the appropriate investigations, in a non-instrumental spirit, as soon as a possible danger is glimpsed.

Furthermore, the knowledge required is not just 'transdisciplinary'. Post-modern lifeworld risks are never just 'scientific'; they are also irreducibly 'local'. The information needed to assess them—not to say to act to lessen them and to deal with the consequences if the worst happens—must also include a great deal that is specific to their particular contexts. As we have seen, it is no trivial task to combine communications from such disparate sources into a coherent body of knowledge. Thus, for example, a thorough analysis of the dangers to health of smoking would not only involve reliable general accounts of its biomedical effects and of the socio-economics of the tobacco industry: it would also include much colourful detail about its cultural significance in various parts of the world and about the problems of securing compliance to legislation. It needs to be controlled in those far-away places. Somehow or another, 'facts' and 'theories' have to be melded with 'narratives' and 'values' to get meaningful results. This applies not only to life-world risks. As we shall see, the post-modern social order as a whole, depends on the systematic production of 'hybrid' knowledge of these very varied kinds.

The post-modern social order

The lifeworld is what individuals experience as conscious social beings in certain places at certain times. They do not, however, suffer the imperatives of society passively. They find niches for themselves in it, adapt its local features to their needs and exert such influence as they can on its larger structures. Indeed, the feasibility of not only comprehending but of reshaping such larger structures — political, legal, economic, etc. — is one of the basic principles of modernism.

The post-modern critique deconstructs this principle. Social phenomena are deemed to be so complex and non-linear that they are essentially unpredictable. It could be said that chaos necessarily rules — whether or not that's OK. That is a naive misconception of the properties of complex systems.[18] Furthermore, it is just simply not the case. Of course our wicked world suffers from innumerable little local difficulties and not a few great big ones. Crime and terror, war and revolution, rapine, disease, famine and poverty, tragically rupture its intricately interlaced bonds. But these disjunctures stand out against a remarkably orderly background.

Science is just one of the things that makes this order possible. Defined narrowly, it is the knowledge base of our material technologies. It thus generates and sustains the multiplicity of communication networks that enable people to coordinate their actions over large stretches of space and time. More generally, it includes the codified legal, political, administrative, and economic knowledge that we rely on in all such coordination. Science is not just a social institution. It is the invisible glue that holds together a great many of our other institutions.

Take, for example, a modern insurance company. Science-based technologies are obviously vital to its operations. Computers, telephones, faxes and emails have replaced the rows of clerks adding up figures and communicating with policyholders by handwritten letters. Sophis-

---

[18] Cohen, J. and I. Stewart (1994). *The Collapse of Chaos: Discovering Simplicity in a Complex World*. Harmondsworth, Penguin.

ticated statistical and financial algorithms automatically calculate premiums and pay outs, investments and foreign exchange transactions. A call-centre in far away Bangalore is optically-cabled to deal with customer complaints in Britain. And so on.

But don't forget that insurance is basically a social artefact. In its modern form it was conceived in the seventeenth century by social technologists seeking to collectivise 'scientifically' the individual risks to life and property. It is stabilised by the 'scientific' rationality of its internal organisation and of the statutes and legal decisions that regulate it externally. Sometimes, alas, insurance companies go bust. But without them, the post-modern world would indeed be totally chaotic.

Modern society developed as a dynamic mosaic of just such institutions. They use science-based technologies and are run on 'scientific' principles. They foster technoscience and benefit from its innovations. They are linked by advanced information and communication systems, and interlock commercially and financially. On the face of it, nothing would seem simpler than uniting them into a coherent system.

As we have seen, this was the technocratic dream. It should not have needed the nightmare experiences of the twentieth century to discredit it. It assigns to scientific knowledge superhuman powers of understanding and prediction. As post-modern critics point out, it overlooks the opaque, fluid, ambiguous and self-referential features of real life that can be neither 'mapped' nor 'modelled'.[19]

Nevertheless, these quasi-scientific institutions — the governmental bureaux, the industrial firms, the legal authorities, the financial corporations, the communication, transport and travel networks, the mass media, the schools, universities, hospitals and armies — are still with us. They are more tightly interlocked and interdependent than ever. These are the 'larger structures' that still frame our life- worlds.

---

[19]   Nowotny, H., P. Scott, et al. (2001). *Re-Thinking Science: Knowledge and the Public in an Age of Uncertainty.* Cambridge, Polity Press.

Up to a point, then, the function of science in contemporary society is simply to rationalise and improve the operations of these diverse institutions. It is called upon to streamline official procedures, create profitable innovations, capture villains forensically, facilitate investment decisions, speed up and broaden out electronic links, design faster, bigger, cheaper vehicles, and so on. Technoscience has its work cut out to meet all these instrumental needs. But the demand for it ought to ensure that an adequate supply of the requisite knowledge is produced

But as we have seen, this conventional view overlooks the non-instrumental roles that science also performs. These are not limited to elementary life-world contexts. Many of them — take the debate over the ethics of human cloning, for example — have to be performed on a much larger social stage. But that stage is set within a culture that no longer tries to define itself in 'modern' 'scientific' terms. So we need to explore the post-modern set-up from this more general point of view. Does our society have any overall order, and how does that relate to science as one of its institutions?

A 'pillar of the state'

The 'modern' concept of governance begins and ends with the nation-state. But in the post-modern social order, so we are told, this monstrous institution has lost its absolute dominance. Well, that may be so, but it hasn't 'withered away' as the Marxists predicted. Nor have the neo-conservatives succeeded in getting it to minimise itself by starving itself of its tax income — financial sustenance they themselves demand for their (national?) military adventures. Regional and global politics stubbornly resist being transformed from the 'international' to the 'transnational' plane. The state is the castle that all political parties seek to *capture*, not to *demolish*. The erosion of the white cliffs of the public domain into the shifting sands of the market place releases the powers of exploitation and corruption, every bit as much as much as enterprise and efficiency.

Nation-states and their subsidiary institutions come in many shapes and colours. They are developing ever more fascinating new forms. Nevertheless they are still much the most important 'large structures' supporting the social order. Think of those unhappy countries where the networks of law and order, commerce and industry, health and safety have broken down. We describe them in two words: they are *failed states*!

The nation-state has always had a warm place for Science. From its origins in seventeenth-century Europe, it has claimed scientific rationality as its organising principle.[20] It relies on technoscience for its unrivaled facilities, for communication, command and control over large populations. Our stock metaphor for its system of governance is no longer of an organism but of a machine. Its working principle is the elementary scientific logic of cause and effect. That is how we describe for example, the transmission of instructions down through the hierarchy of a state bureaucracy or army. That looks like the 'scientific' way to ensure order in society at large.

In reality, of course, what we blithely refer to as the 'apparatus' of government never works as smoothly as even the most primitive steam engine or cuckoo clock. The postmodern political theory amplifies the frictions and glitches, and throws doubt on the rationality of the whole process. In an open society, 'mafias' and 'sleaze' can be publicly denounced as deviations from a normally equitable and predictable social order. In a technocratic autocracy however they inevitably go on growing until they completely gum up the works.

So the post-modern state doesn't pretend to 'do' control. But it goes in quite frenetically for *regulation*. As a rational institution it tries to use rational means to restrain the irrational forces — greed, prejudice, cruelty, vanity, rivalry and suchlike human propensities — that challenge the order that it tries to sustain. And it fears above all the threats

[20]   Ezrahi, Y. (1990). *The Descent of Icarus: Science and the Transformation of Contemporary Democracy.* Cambridge MA, Harvard UP.

posed by the *military* capabilities of rival states. These it can neither control nor regulate, but it has very good reason to *defend* itself against them.

For all these purposes, technoscience is clearly the ideal instrument. As we have seen, nation-states support it in this role on an enormous scale. The next Martian science tourist who asks: 'Take me to your leader' should be directed towards the Pentagon. There, if anywhere, is the principal stronghold of science in our globalised society.[21]

For example, debates on science policy tend to skim quickly over the large entries for military technoscience in national budgets. Or, rather, these expenditures are accepted as subsidiary items in the overall cost of acquiring the specific capabilities they promise to enable. Political controversy can erupt over, say, a proposal to spend untold billions on an unnecessary fleet of new fighter aircraft. Yet very little attention is given to the fact that as much as 50% of this vast sum will have to go into Research, Development, Design and Demonstration — 'RDD&D' — before a single aircraft takes off from the assembly line.

In effect, this mode of knowledge production is fully incorporated into the post-modern state system. Its deep-seated scientific uncertainties and ethical ambivalences emerge as 'technical difficulties' or 'managerial problems' and 'political questions'. Take, for example, the National Missile Defense project being pursued by the US Federal Government. Outside that country, this is a subject of highly critical debate. But the focus is on its effect on the international *status quo*. There is almost no discussion of genuine scientific doubts about the technical feasibility of constructing any such new system, nor of the deceitful ways of American officialdom in covering up their failure to move convincingly towards this goal.[22]

The place of science in society is thus deeply coloured by its intimacy with the war-making facilities of the nation-

[21] Salomon, J.-J. (2001). *Le scientifique et le guerrier*. Paris, Belin.
[22] Mitchell, G. R. (2000). *Strategic Deception: Rhetoric, Science and Politics in Missile Defence Advocacy*. East Lancing, MI, Michigan State University Press.

state. This obviously has a profound effect on how knowledge is produced and disclosed, right through the research system. The partnership with these agencies of the state is further complicated by close ties with industrial corporations and academic institutions. As we shall see, one of the major social roles of non-instrumental science is to counterbalance the technoscience in the hand of this many-bodied complex.

The state as patron

As we have just noted, the rationale of the nation-state obliges it to support regulatory research on a similarly large scale. This logic naturally extends to the science required to sustain and improve various more positive public benefits, such as good schooling, reliable weather forecasting or the control of infectious diseases. These shade further into much more general, longer-term, 'strategic' goals, such as curing cancer, preventing inflation, maintaining economic growth, facing up to climate change, and so on.

National governments, legislatures, and the voters that elect them are evidently very willing to support science for such benevolent purposes. Throughout the twentieth century, this responsibility was increasingly taken on by the state. It has not been seriously weakened, either by postmodern political scepticism or by the neo-conservative demand for 'market solutions'.

In principle, the knowledge thus produced — for example by state-owned biomedical research organisations such as the NIH in the USA, the MRC in the UK, INSERM in France. — is for societal benefit and public use. In raw form, it is seldom capable of being assimilated by ordinary people. But it is freely available to the professional practitioners and commercial enterprises who cook it up and distribute it for lifeworld application. In the past, the transfer of knowledge along this chain was often impeded by institutional walls and moats. One of the beneficial features of post-academic science is to level these barriers. If, say, researchers in the MRC's Laboratory of Molecular Biology make a discovery that really opens the way to a cure for cancer, the appropriate drug will fairly soon be prescribed by physicians and stocked by pharmacists. So this apparently distant state-owned mode of knowledge production is actually possible to connect quite effectively into our society through a variety of well-established channels.

I am not arguing, of course, that this is necessarily socially beneficial. For example, there is much controversy about the true value of 'conventional' medicine. The plain fact is that the great majority of people think that it is, and assign an important social role to state-funded science designed to support it. In essence, public money is being used to produce a public good. Although it is somewhat fanciful to try to evaluate this in terms of economic costs and benefits, such exercises typically show that it is a vastly profitable investment for society as a whole.[23] As we have seen, this is the basic reason why 'post-academic' science is funded so generously from the public purse.

The real trouble is that the post-modern political practice of 'government funding' for research is easily confounded with an older tradition of 'state patronage'.[24] The history of science reveals a litany of petitions by scientists for the means to undertake their investigations. Emperors, kings, princes and ministers of finance have always seemed the richest targets and softest touches for such appeals. Usually they have been promised untold benefits for themselves or for their subjects if they shell out a few ducats, guineas, or dollars. And usually the would-be inventor has not been held to account if his project hasn't—as it almost always couldn't have—actually produced any tangible profits.

More carefully, the savants put it around that the production of knowledge was bound to be advantageous to the land that fostered it. Even if the riches thus acquired could not be realised for several generations, they were not stored entirely in heaven. The generous patron of science was an

[23] Mansfield, E. (1991). 'The social rate of return from academic research.' *Research Policy.*

[24] Turner, S. P. (1990). Forms of patronage. *Theories of Science in Society.* S. E. Cozzens and T. F. Gieryn. Bloomington IN, Indiana UP: 185-211.

intelligent philanthropist, who gained honour by making gifts whose value actually increased with time.

Well, that's a grossly oversimplified interpretation of a fascinating and complicated historical phenomenon. But until the twentieth century, state patronage of science was typically *ad hoc* and sporadic. Academic science was primarily the financial responsibility of the institutions where it was located. Universities mostly had to exist on their teaching income and the support of their alumni. Even in those institutions that were organs of the state, and/or heavily subsidised from the public purse, the professors were expected to compete for their research resources from limited institutional budgets.

Various historical circumstances and trends began to put quite impossible demands on these sources. Academic science simply could not survive without systematic public funding. The modern rationale for this policy was famously expressed by Vannevar Bush in 1945.[25] As we have seen, this report argued 'strategically' for the 'pre-instrumental' benefits to the USA of the unprogrammed search for knowledge. Other nation-states used rather similar arguments to justify their 'no strings' patronage of academic research. Nowadays, it is assumed without debate that this is where science gets a major part of the income it needs to maintain its accustomed place in society.

The academic scientists who receive this largesse are duly grateful for the tribute to their personal commitments and material interests. It is fair to say, though, that they, and many of the enlightened politicians and officials who hand out the funds, also appreciate this policy for more idealistic reasons. Indeed, for them 'academic freedom', 'the honest search for truth', 'the cultural value of pure science' (etc.) are absolute social virtues. Their actual benefit to society is not to be questioned or explained. For example, I once tried to persuade the top brass of the French CNRS (National

---

[25]   Bush, V. (1945). *Science – The Endless Frontier: A Report to the President on a Program for Postwar Scientific Research*. Washington DC, National Science Foundation.

Centre for Scientific Research) to give their corporate plan an explicitly 'strategic' rationale. But they rejected the notion on the grounds that the academic research being done by their ten thousand or so extremely talented scientists at state expense should be explicitly justified as 'pre-instrumental', whilst they just took for granted its more general 'non-instrumental' social function.

Post-academic science is thus firmly upheld by the state. This is an indication of its high standing in the post-modern social and financial order. But its place is ambivalent. On the one hand, it receives generous state subsidies as an independent institution making unspecified contributions to the life of the nation: on the other hand, it is funded by the government for the evident services it renders, directly and indirectly, to the economy and the polity.

The hypothetical 'contract of science with society' is really only a tacit agreement between the research community and the state. You give us a pile of unmarked bank notes, and we will provide you with a mixed bag of goods, some of which will be very valuable! But its two main clauses are contradictory. The terms of the first clause forbid the state to interfere in any way in the knowledge production process. By contrast, the second clause gives it the power and motivation to supervise this process and direct it towards its own purposes.

The balance between these two policy principles has always been unstable. In some countries, such as France, it has been maintained by keeping purely academic science quite separate from governmental and commercial technoscience. In other countries, such as the UK and the USA, awkward questions have been avoided by deliberately smudging this distinction. But the insecurity of the state itself, under post-modern conditions, drives it towards the 'safety' of accountability and control.[26] As we have seen, in post-academic science even the most basic

---

[26] Cozzens, S. E., P. Healey, et al., Eds. (1990). *The Research System in Transition.* Dordrecht, Kluwer, Ziman, J. M. (1994). *Prometheus Bound: Science in a Dynamic Steady State.* Cambridge, Cambridge UP.

research project is supposed to be able to produce results that could conceivably be exploited practically.

Does this shift towards instrumentalism matter? The 'State', after all, is only a name for the complex of institutions by which society is officially kept in order. Under normal conditions, these institutions follow the instructions of the government. In a democratic country, the government is run by the people, for the people. So state-managed science is necessarily of benefit to society. Its social role is scrupulously scripted. It can always be trusted to be on the side of the angels. 'I am your duly elected Minister of Food and Agriculture, and I am advised by my scientists that it is perfectly safe for you all to eat as much of our excellent British beef as you can stomach. ('Look, even my darling daughter is wolfing down a beefburger!')

Alas; neither democracy nor science are so perfect. They are both of them pluralistic, contentious and confused. A high proportion of the court cases involving science are between an agency of the state and a private citizen or corporation. Ethical concerns about the applications of technoscience — for example, therapeutic human cloning — often become party politics. Governing groups, and the state machinery that they control, are easily influenced by other centres of social power. In fact, as we have seen, many of the most important non-instrumental functions of science require as much independence from the government of the day, as from any commercial or industrial interests.

Furthermore, the administrative apparatus of a democratic state is not well suited to the management of scientific research. Its bureaucratic norms favour routine and reject risk. Its accent on accountability cannot be harmonised with the uncertain promises of experimentation. Having to be answerable to its political masters, a publicly-funded research system is vulnerable to the whims — and also to the not so whimsical material interests — of legislators. In practice, elaborate systems of committees, councils and codes are required to buffer its budgets and shield its operations from direct political interference.

As we have seen, the post-academic mode of knowledge production has evolved within this environment. It has developed various systematic practices, such as the peer review of projects that keep it from being entirely subservient to the state. Corporate and foundation funding gives it some independence from overtly political influences. But its official and commercial connections shut it off from the countervailing forces that keep the post-modern social order from collapsing into technocratic tyranny. As we shall see, it is only by opening these channels and aligning itself morally with 'civil society' that science can perform an enlightened social role.

Science in the marketplace

Another sweeping generalisation about the post-modern condition is that it is dominated by 'the market'. This is not only supposed to have taken over from 'the state', it is also treated as if it were a coherent institution, a source of forces capable of blindly crushing all who oppose it. This uncontrolled steam-roller metaphor serves well those who are adept at pushing weaker opponents into its path—the CEOs of monopolistic multinational corporations, for example. But what we are really talking about is a bundle of contrived social practices systematically sustained by the authority of the state. The post-modern social order very strongly favours market relationships, such as the use of a stable currency to mediate the regular trading of goods and services amongst individuals, and business firms.

It is true that what is called 'economic rationality' increasingly pervades the public domain and percolates into the more personal areas of the lifeworld. The traditional boundaries are breaking down between 'producers' and 'consumers', between 'sub-contractors' and 'manufacturers', between 'distributors' and 'retailers'. World-wide negotiations are conducted at electronic speed. Stock Market options, refinancing, outsourcing, and other devices confuse the contracts that bind people together into free-market relationships. But only utopian, hyper-libertarian, neo-classical economists imagine that society could be held together solely by a web of such obligations.

It is instructive, nevertheless, to look at the production of scientific knowledge from that perspective.[27] As regards technoscience, this is perfectly straightforward. Scientists sell their technical services to industrial firms. With the help of these services, firms invent novel products and processes. The knowledge required to reproduce these artefacts is deemed by law to be 'intellectual property'. As such it can be sold or leased to other firms, either as a trade secret or as a legally protected patent.

An item of technoscientific knowledge is thus a marketable commodity. It is not, of course, as tangible as a cargo of wheat or even a suitcase full of heroin. Its value depends enormously on the context, and may vanish overnight if a better mousetrap is invented. But it stands up pretty well in a world where such insubstantial entities as company logos, commercial goodwill and managerial reputation all have their price.

Academic science is quite different. The knowledge it produces is not for sale: it is for *free*. In truth, the norms and practices involved in its production have 'economic' analogies. Contributions to the scientific literature are 'exchanged' for recognition by citation.[28] Research claims are put on display, and critically selected for 'purchase' and use by other researchers.[29] Scholars compete for employment in the academic 'marketplace',[30] and so on. But these are mere metaphors. What comes out of the process is what may be called a public good. It does not belong legally to either the

[27]   Ziman, J. M. (1991). 'Academic science as a system of markets.' *Higher Education Quarterly* 45: 41-61, Ziman, J. (2002). The microeconomics of academic science. *Science Bought and Sold: Essays in the Economics of Science*. P. Mirowski and E.-M. Sent. Chicago, University of Chicago Press: 318-40.

[28]   Hagstrom, W. O. (1965). *The Scientific Community*. New York NY, Basic Books.

[29]   Ziman, J. M. (1968). *Public Knowledge: The Social Dimension of Science*. Cambridge, Cambridge UP.

[30]   Caplow, T. and R. J. McGee (1958). *The Academic Marketplace*. New York NY, Basic Books.

people or the institutions that have generated it: neither can it be appropriated by those who wish to exploit it.

Of course, as we have seen, the scientific knowledge in the public archives does have enormous pre-instrumental potentialities. In the long run, some of it will prove to have created immense wealth, both for a few individuals and for society at large. The eventual payback for public expenditure on its production dwarfs the return on any other form of investment. But this profit is highly speculative in detail, and cannot easily be recaptured by those who financed the research originally.

What is more, many of the social benefits of academic science are genuinely non-instrumental. As we have seen, it can help society to run more smoothly, anticipate distant perils, and give people more fulfilling lives. These benefits are perfectly real and absolutely essential for the maintenance of the social order. But it is nonsensical to try to quantify them in monetary terms.

So the judgement of 'the market' on post-academic science is always dangerously one-sided. It is not just that its residual non-instrumental features are not amenable to economic rationality: they simply do not exist on the economic map. Every managerial drive to improve the efficiency of research organisations treats them as 'waste' and seeks to eliminate them.

Here is the sort of thing that happens. Imagine a medium-sized scientific institution, of excellent standing, founded in the early years of the twentieth century. I really don't have any specific case in mind, but to give some idea of the type of research that it might have undertaken let me call it the National Institute for Bio-Agronomy — 'NIBA'. In other words, it was mainly devoted to the production of basic scientific knowledge, with rather broad strategic implications for a major sector of the economy.

Anyway, until the 1970s NIBA was supported by the state for all its scientific work. Its income came through one of the governmental science funding bodies in the form of a block grant, revised in amount and broad purpose every five years. Its scientific staff were tenured civil servants, but many also held honorary academic posts in the neighbouring university. There was never any dispute about the long-term economic importance of their published work. Even though this didn't always relate directly to the immediate problems of the relevant industries, they were in regular demand as consultants and advisers in both the public and private sectors. In essence, they were engaged in 'academic' science. That is, they worked as individuals or in small groups on problems that were largely of their own choosing, and gained personal recognition — typically, promotion to higher official rank — for their published contributions to knowledge.

It then began to be felt, in higher political circles, that NIBA ought to give more attention to 'national needs'. What this really meant was that its research programme should be more directly decided by the government. So it was decreed that a proportion of its income should now come directly from various government departments, through contracts to do research on explicit topics. This rather clumsy procedure was chosen, no doubt, so that 'state rationality' should not be seen to prevail too obviously over 'academic freedom'. But it was also deliberately modelled on the arrangements by which large companies often 'outsourced' some of their research and development requirements. It thus opened the way for NIBA itself to seek funds from the private sector for similar work.

This certainly had the desirable effect of forcing NIBA to be thoroughly up to the minute in its research capabilities. Its scientists became skilled in the use of electronic instrumentation, supercomputing, genetic engineering and other advanced (and expensive) technical procedures. They were also forced to work more closely together in research teams, and to undertake projects arising in 'contexts of application'. In effect, although they were not fully engaged in 'technoscience', their system of knowledge production soon acquired many 'Mode 2' features.

In addition, it opened a space for the exercise of 'economic rationality'. Governments had to make big expenditure cuts, so NIBA was advised to earn more of its income

from commercial contracts. Being highly expert as researchers, and more enterprising than their 'civil servant boffin'
stereotype suggested, its scientists took up the challenge
very successfully. By the beginning of the '90s, it was mostly
operating in that mode.

At that point, however, it seemed feasible to follow the
dictates of the market ideology and to 'privatise' NIBA completely. On paper, at least, it would seem capable of operating as an independent enterprise. It ought to be able to
survive comfortably as a 'research boutique', selling its specialised services to larger corporations and state agencies.
After some 'regrettable' staff redundancies, it was reconfigured managerially in that form.

Unfortunately, as we have seen, this corner of the business world is very unstable and financially insecure. Within
a few years, NIBA had run out of capital, and had to be
'rescued' by one of the large multinational firms for which it
had been working. It was taken over as a going concern, at
first as the 'bio-agronomic research centre' of the firm but
soon incorporated into its normal industrial and commercial structures. More of its staff were 'let go', but the best of
them are now expertly—and profitably—engaged in very
promising technoscience projects likely to result in some
highly competitive technological innovations.

So that's all right, isn't it? But just this year a great public
controversy blew up over (let us say; since this is all entirely
fictitious) the safety of genetically modified foods. The government and the public looked around for scientists capable
of doing the extremely specialised research required to
throw more light on the problem. Formerly, this could have
been carried out by NIBA, or by members of its staff acting
as independent consultants. Now they are all employees of
one of the very firms whose products are in dispute.

The details of this fable would perhaps have read differently if it had been written in another language, such as
French or Swedish. But the moral would be the same.
Post-academic science does not yet have a stable situation in
the post-modern social order. For all economic, cultural and
political influence, it is not a self-contained institution, nor a

network of institutions, like the banking system or the railways. It has vacated its Ivory Towers, but it is not reconfigured to stand alone and its tastes are too costly for it to live by its own wits. As we shall see, much of it still co-habits with higher education, but with waning enthusiasm on either side. Financially, it is suspended uneasily between the state and the market, each of which makes much use of the knowledge it produces. But neither of these has a vital interest in maintaining the features that enable science to perform its traditional non-instrumental social role. As I shall argue in the final chapter, this challenges the Third Force in our pluralistic culture—'civil society'—with a responsibility which it should not decline to support.

The global setting

Before we drop back down to the grassroots—another tree-top cliché—we may consider the now familiar ways in which the post-modern culture is *global*. It transcends geopolitical frontiers. Companies are typically *multinational* Electronic money flows instantaneously round the earth. *International* agreements regulate our comings and goings. Goods are traded along supply chains that encircle the planet. We ourselves inhabit cosmopolitan life-worlds with their own everyday know-how, and nation-states have ceded away many of their sovereign powers to *supernational* organisations.

So of course, science has gone global too. It has always been the least parochial of all cultural forms. It endeavours to produce *universal* knowledge. That applies more to the natural sciences than the human sciences, but even history cannot resist being trimmed to fit national boundaries. It was always the boast of European civilization that its scientists and scholars communicated and travelled without regard to nationality.

Of course they had to perform their duties as loyal subjects or citizens of their native countries. They were not immune from sectarian prejudices. They frequently devoted their knowledge and skills to war-making, and occasionally to ultranationalist or racist polemics. But the community was always, in regular practice as well as in celebratory principle, a transnational 'republic of learning'. This applied not only to 'academic' science. Technoscience had always traveled in company with the trade in its artifacts, alongside silk, spices and new weapons of warfare. Stories about rulers who attempted — usually unsuccessfully — to prevent the export of their technological secrets only show how globally fertile these ideas were thought, quite rightly, to be.

That is not just a romantic ideology. The important point is that there are no deeply entrenched obstacles to the post-modern 'globalisation' of the production, distribution and use of scientific knowledge. A multinational research team is not that much more difficult to manage than a multidisciplinary one: at least everyone speaks the same technical jargon. In fields such as ecology, international research projects often make more sense than purely national ones. The grant-seeking strategies required to get funds from intergovernmental bodies such as the European Union are not very different from those learnt in the national leagues of the same game. Intercontinental air travel to a conference need take scarcely more time than a cross-country railway journey, and is paid for by the parent institution. And a telephone conversation or an email exchange with a far-distant scientific 'peer' may be easier to set up than a link with a busy colleague in a building just across the campus.

One might say, rather, that it is Society that has changed more than Science. The globalisation of the social order has made it generally more receptive to scientific ideas and artefacts. This, of course, was one of the major goals of modernisation. For a country of the 'South' to develop economically and progress socially, its traditional culture surely had to give way to 'Western' science and technology.

To a considerable extent, this is happening. Recent products of technoscience, such as the contraceptive pill, the cell phone and the AK47, are breaking the mould in every village. Local land-owning elites are building factories and doing business on the internet. Students of science, technology, engineering and medicine crowd into the universities in every city and go abroad for advanced study — and employment. Governments hastily set up research institutes to foster competitive innovation. Multinational companies invest in mines and plantations to extract and export, with scientific efficiency, the natural products and commercial crops of the region. And so on, to the tune of trillions of dollars in currency transactions, billions in the flow of sophisticated material goods and intellectual property, and millions of 'technical migrants' gaining knowledge and carrying it home inside their heads.

Unfortunately, the last quarter of the twentieth century exposed the moral vacuum within the modernisation project. Its technocratic imperatives tore apart the ecological, economic, ethical and ethnic webs that sustained other environments, lifeworlds and social orders. Perhaps that was unavoidable. But post-modern society is in such a mess globally that it has no distinct place for science.

Take governmental institutions, for example. As we have seen, a substantial proportion of post-academic science is still undertaken by national governments, and a lot more of it is heavily dependent on public funds. Globalisation means that nation-states hand over some of their regulatory powers to intergovernmental bodies such as the European Union. This transfer ought to include financial and managerial responsibility for the corresponding research. But this frequently turns out to be more bureaucratic and politicised, and less well supervised and funded, than the scientific work performed by traditional state institutions. Furthermore, the research on health, food, agriculture, etc. undertaken by international organisations is not nearly sufficient to serve the needs of a world where many post-colonial states are so poor, weak and disordered that they cannot maintain competent scientific institutions of their own.

Strangely enough, some of the 'academic' components of post-academic science have globalised themselves more effectively than their more instrumental partners. It is true that they get little practical help out of UNESCO — the organ of the United Nations designated to serve education, science and culture. And they only get a few grudging crumbs from general intergovernmental organisations such as the EU or the World Bank. But, as we have noted, academic scientists are already transnational in spirit. Through their inter-personal and inter-institutional networks they have induced national governments to fund jointly a number of international facilities, projects, programmes and institutes. CERN, for example, with its immense particle accelerators, is financed and managed as a single unit by an intergovernmental consortium, whilst the mapping of the human genome was undertaken as a loose collaboration of public, academic, charitable and private research institutions in many countries.

Notice, however, that the production of scientific knowledge is not a self-contained enterprise. It is a long time since it was the work of solitary individuals. Nowadays it is seldom undertaken by independent social units such as small firms. Scientists build their nests in well-established organisational niches, or amongst the branches of stable institutional forms. For at least a century, many of these nesting sites have been provided by the state. Post-academic science is thus culturally adapted to the modern social order, where its high place was assured by a strong nation-state. In the post-modern world, where there is no predominant global authority, this place is ill-defined and much less secure.

What really does operate globally is the Market. This means that money, goods and services have much more freedom to be traded around the world. The main consequence for Science is the emergence of larger and larger commercial corporations. Multinational firms such as General Motors and Unilever exceed many nation-states in population, wealth and influence. And they cultivate technoscience on a correspondingly gargantuan scale. The post-modern social order is based on the products of their cosmopolitan laboratories and transnational research systems. Here, surely, science has an honoured place.

And yet it is doubtful whether this place is more secure or weighty than it was a quarter of a century earlier. The research and development divisions of the major industrial firms were quite large enough to invent, test and bring to the market all the novel artefacts people could afford to buy. Competitive technological innovation was quite powerful enough to drive the capitalist economy at its accustomed breakneck rate. There were jobs enough for all the 'QSEs' — Qualified Scientists and Engineers, as they are classed in the official statistics — who poured out of the universities and colleges. Outstanding technical achievements were widely celebrated and well rewarded. Scientists complained, of course, that they ought to have more control of the companies that employed them: as it turned out, this may have been the high tide of their influence in the commercial world.

The trouble is that free market systems are driven by the promise of short-term profits rather than by technical rationality or public welfare. Companies that are already very large have to go on growing, not so much in the vain hope of achieving further economies of scale but to avoid being taken over by yet bigger companies with deeper pockets. When this happens, perfectly productive and viable research and development operations are broken up, merged with others, shifted around the world or otherwise reconfigured. So personal careers that once looked very stable have become fragmented and insecure.

On the other hand, as we have seen, post-academic science is being transformed by a number of other marketplace developments. These include the rediscovered vigour of small inventive enterprises, the commercial advantages of 'outsourcing' research and development to independent subcontractors, the surprising successes of 'open source' software development, the post-industrial deconstruction of managerial hierarchies and the deployment of researchers in temporary, multidisciplinary project teams. Jobs for 'QSEs' in a globalised high tech industry have certainly become more risky, but the prizes of success are bigger. They are also more diversified and less rigidly specialised, more sociable and less autonomous. And there are often openings for the same type of work in cosmopolitan research centres in many different countries.

Nevertheless, as the firms with the largest slices of the market consolidate their position, they concentrate their technoscience into a smaller number of large projects. In the absence of global regulatory authorities to prevent further mergers and acquisitions, the laws of free market competition decree that oligopoly should give way to a what is effectively a monopoly. After all, if competition is serious, its winners must grow and its losers shrink. But a substantial market share is already a big financial advantage in any such contest. In the long run, one winner takes all the business and the others have to kow-tow to it or go bust.

Thus, 'economic rationality' leads back to the spurious rationality of technocracy. Technological evolution by commercially competitive innovation[31] comes to be thought 'wasteful'. Ultimately, everybody is inside one big tent producing carefully licensed intellectual property—and comfortably profitable goods for appropriately conditioned consumers—to the specifications of an unchallengeable corporate bureaucracy.

At present, fortunately, this is only science fiction. The global market leaders in most industries still compete ferociously at the forefront of technoscience. In any case, we must not ignore the creative sparks that so often set fire to the technological underbrush of our industrial culture. Corporate R&D laboratories do not always originate the discoveries and inventions that they have developed for the market. Practical innovations still come from individuals or very small enterprises financed precariously with speculative capital. Knowledge is thus being produced by a whole variety of people who are largely self-taught—skilled technicians, self-employed craftsmen, proprietors of small workshops, school teachers and other members of a technological 'lower middle class'. One of the desirable features of

---

[31]   Ziman, J. M., Ed. (2000). *Technological Innovation as an Evolutionary Process*. Cambridge, Cambridge UP.

an open, democratic society is the cultivation of this type of 'folk technoscience'. The essence of a truly free market is that it provides the opportunities and means for such enterprises to germinate and to attempt thrive.

Nevertheless, the economic globalisation of techno-science has distanced it further from its contexts of application. A large multinational company typically performs its RDD&D (Research, Development, Design and Demonstration) operations in a small number of centres located in the 'industrial parks' of North America, Europe and Japan. The scientists, engineers and technologists in these centres are linked to one another globally, but have few direct connections with the diverse lifeworlds where their artefacts will be put to use. Local knowledge derived from practical experience has to be transferred through many network nodes, up the bureaucratic hierarchies of several subdivisions of the corporation, before it finally filters down to the research bench or computer database.

This disconnection between technological innovation and practical use is having disastrous effects in the Third World. Many developing countries lack the technical knowhow, financial capital, patent rights, industrial base and profitable markets to manufacture high-tech goods adapted to their particular needs. In some cases, people are induced to buy, at exorbitant prices, products that have apparently been designed with complete disregard for the way they will be used in practice. In other cases, they are not even offered elementary artefacts, such as cheap generic medicines and simple agricultural machinery, appropriate to their actual lifeworlds. In the global market place, the hidden hand of profit often fails to foster the pluralism and utilitarianism that are supposed to be its supreme virtues.

Knowledge as power

Knowledge enables meaningful *action*. Action is the means by which *power* is exerted. Thus, knowledge is an essential pre-requisite for the use of power. Science is a social practice producing knowledge. So power can be gained by the cultivation of science and powerful institutions defend themselves by controlling science and appropriating its output. The agenda of this book is really very simple. To discover the place of science in society, locate the centres of power. Alternatively, find out where the science is, and there you will find the power. *Politics*, with an *economics* bodyguard, is big in this investigation.

But how does this elementary social logic manifest itself in the post-modern world? As we have seen, we already live in an *information* society. We are wired for more sights, sounds and text messages than our minds can receive. The internet opens to us millions of pages of archival records and current news. In many countries, access to governmental information is the legal right of every citizen. Science is for public understanding and the public good. Ours we tend to think is a truly democratic society, where knowledge is just as widely distributed as power.

In reality, however, the situation is rather different. Post-modern society is *pluralistic*. It is a network with nodes where power is heavily concentrated. Various highly organised institutions, some under the direct authority of the state, others sustained by market forces, some national, some global, compete for influence. In this arena, private knowledge is increasingly the decisive weapon.

But technical knowledge is difficult to conceal for long. Even that most dreadful of state secrets, the recipe for an atomic bomb, was ferreted out of Los Alamos by Soviet spies much sooner than anyone expected. Furthermore, technoscientific progress continually changes the field of battle. Nowadays, it is the design of gas centrifuges and of electronic countermeasures against nuclear missiles that must at all costs be kept out of the hands of 'rogue states'. And in commercial arenas, the knowhow embodied in novel artefacts is quickly found out, and can only be protected for a while by intellectual property law.

Powerful institutions are thus compelled to be continually active in science and technology. To maintain their competitive position, in military strength as well as the peaceful arts, they have to be systematically engaged in the production of new knowledge — that is, in *research*.

In principle, this is no more than is advised by the 'scientific attitude'. As we have seen, one of the non-instrumental functions of science is to promote 'evidence-based' decision-making. In spite of philosophical scepticism about the effectiveness of 'rational' action, this is a widespread feature of the post-modern culture.[32] Look at the formal findings of any commission of inquiry, or 'working party'. In the 'executive summary', the recommendations for action are referred back to the data presented in the main body of the report and set out at length in a number of technical annexes. Although these data usually exist already in one form or another, their collection and analysis may have required a systematic literature search or public survey — which is very often the only lasting achievement of the report!

This form of research is, or appears to be, open to everyone. But a large organisation can afford to take it much further. The latest data-processing equipment is very expensive. But computerised data collection, surveillance, pattern recognition, number crunching, search algorithms, etc. can provide significant information that is completely invisible to the average punter. Military intelligence, crime prevention, financial analysis, market trends, consumer preferences, voting intentions, etc. are all typical contexts of application for 'Mode 2' technoscience. They are one of the post-modern ways in which science reinforces power with knowledge.

But these processes mainly reveal existing information, or indicate how it is being recombined or reconfigured. Governments and large industrial firms are fighting for their lives. They have to have privileged access to truly new facts and ideas. Sometimes, as we have seen, the necessary intellectual property is for sale or rent. But good in-house facilities are always needed to evaluate and utilise it. So the

---

[32] Nowotny, H., P. Scott, et al. (2001). *Re-Thinking Science: Knowledge and the Public in an Age of Uncertainty.* Cambridge, Polity Press.

only safe option is to invest in the means by which new knowledge is produced. This means that large enterprises are forced to engage actively in normal, laboratory style, instrument intensive, expertly staffed, theory guided, experimentally tested, problem oriented technoscientific research.

This is often a very profitable investment. But it cannot be done on the cheap. As we have seen, good science nowadays is heavily capitalised. It often requires access to a variety of expensive instrumental facilities. Very frequently, a serious research project can only be carried out by dozens of scientists working together as a team. The nearer it gets to the marketplace, the more it costs, project by project. The critical mass of money needed to get useful results may be tens of millions of dollars. It is just a waste of money to spend less. In other words, only an enterprise with immense financial resources can afford to undertake the research it really needs for its own survival.

To put this important point briefly. Existing *knowledge*, open or secret, is always of limited relevance to the unfolding novelties of the post-modern world. So it is not knowledge as such, but *research* capability — the ability to generate new knowledge to order — that is the real key to power. But this capability is now very costly, and only available in indivisible lumps. It is mainly in the hands of very large organisations, whose political and economic dominance it further strengthens.

The prime example is the vital role of technoscience in the 'military-industrial complex'. In 1961, President Eisenhower warned the United States against the growing influence of this conjunction of an immense military establishment and a large arms industry. Nowadays, it is customary to add 'academic' to the duo of 'military- industrial'. Knowledge relevant to the conception, development, manufacture and use of weapons of war flows back and forth between organs of the state, commercial firms and university research laboratories. Little of this knowledge gets into the public domain or on to the open market. What comes out, eventually, is naked power, as from the barrel of a gun.

The close involvement of science with war is an old story. From the earliest times, they have lived and evolved side by side.[33] What is new is their social symbiosis. The half century after Hiroshima has integrated technoscience into the war-making machinery of the state. The advanced nations are spending a significant fraction of their national incomes on R & D for direct military purposes. Multinational companies renowned for the technical quality of their consumer products are also actively engaged in the international trade in arms. It is considered quite normal that something like one quarter of the students being taught science or engineering are going to be doing this work. Leading scientific authorities — not merely the administrative Chiefs but the Intellectuals as well — are personally involved, even in peacetime, with decisions on military technique and strategy. It must be emphasised that the temporary mobilization of the scientific community in the two World Wars has now been made permanent.

The ethical implications of this development are obviously extremely grave.[34] It exposes a moral contradiction at the very heart of science. On the one hand, few people nowadays consider war to be a virtuous undertaking. At best, it is sometimes a dreadful, regrettable necessity. Despite the paradoxes eloquently dramatised, a whole century ago, by George Bernard Shaw in his brilliant play *Major Barbara*, the morality of the arms trade is particularly questionable.

Science, on the other hand, always claimed to be an essentially noble enterprise. Scientists were depicted as 'honest seekers after truth', whose peaceful intentions could not be doubted. Whether or not this reputation was justified, it enabled those who took on the virtues of peace to perform some vital social functions. Anyway, it could truly be said of the great majority of the scientists of pre-war days, that they

[33]   Constant, E. (2000). The evolution of war and technology. *Technological Innovation as an Evolutionary Process.* J. Ziman. Cambridge, Cambridge University Press: 281-98.
[34]   Salomon, J.-J. (2001). *Le scientifique et le guerrier.* Paris, Belin.

were only marginally associated with the war. Academic science, in particular, was almost entirely free from any such connection.

Now, alas, the contradiction between the peace-loving and war-making aspects of science stares us all in the face. Let us leave aside, for the moment, its ethical dimensions. The plain fact is that this connection with modern armaments can no longer be denied. It is disingenuous for scientists these days to dissociate themselves entirely from an activity that engages something like a quarter of their colleagues in highly secret activities aimed at ever more destructive warfare.

It is not so easy, even, to isolate oneself professionally from the all-pervading *research network* of the military-industrial-academic complex. As we have seen, post-academic science customarily plays down the political or commercial goals of the bodies that fund it. As a result, the boundaries between military and non-military research are thoroughly blurred. Should one really refuse a grant from a devolved subsidiary of a ministry of defence to carry out a project that seems purely 'basic'? How could one know that they are actually supporting work on the quantum gravity of spinning bodies because they believe it might facilitate the further development of inertial guidance systems for nuclear missiles? At what point does the investigation of extremely virulent pathogens leave the bureaucratic turf of preventive medicine and become an item on the shopping list of defensive biological warfare?

Notice here that the research is not being controlled with an iron fist. One of the unwritten procedural norms of post-academic science is not to allow funding bodies to dictate the actual results of the projects they finance. They often give themselves the legal right to delay the publication of awkward or potentially exploitable findings. But it is considered scandalous to misrepresent these findings or keep them secret for long. This is a significant constraint on the knowledge-shaping capabilities of even the most powerful organisations. For example, nothing is more discreditable for a pharmaceutical firm than for it to become known that its agents are deliberately suppressing information about the unfavourable outcome of a drug trial. Similarly, the dishonesty of the official reports on tests of missile defence systems puts these grotesque enterprises beyond the bounds of scientific credibility.[35]

In practice, the organisations that support research exert their power more subtly. They cannot prescribe the outcome of a project, but they can decide whether it should have been undertaken at all. This is a normal managerial prerogative for their in-house R&D operations and in their external research contracts. But it also applies to the funding of project proposals, whether unsolicited or in response to a request for funds. In my experience selection is weaker than prescription, but in the long run it shapes the world just as effectively.

It can be argued, of course, that control over the initiation of a research project is not significant, since there is no certainty that it will discover what is promised. But it is far from ineffectual, for a decision not to undertake this particular project surely prevents any such discoveries at all. 'The truth will out', we say complacently, as if some time in the next millennium will do. Any way, we are talking of large programmes and policies, where the fate of individual projects is not the determining factor. Our experience is that a concerted technoscientific attack on a wide front very often reaches its objectives. So the capacity to produce knowledge by this means is a potent adjunct of political and/or economic power.

It is true, as we have seen, that many of the organisations financing post-academic science are quite sincere in their efforts to support 'good science'. They award research grants on the basis of scientific merit, as assessed by peer review. Nevertheless, these bodies often have wider official purposes, such as the enhancement of the economic competitiveness of the nation. These purposes are inevitably

---

[35]   Mitchell, G. R. (2000). *Strategic Deception: Rhetoric, Science and Politics in Missile Defence Advocacy.* East Lancing, MI, Michigan State University Press.

reflected in the selection of projects for support. Even though this criterion is only loosely applied it may generate a significant bias in the knowledge that is eventually produced. Thus, for example, it may favour the technologies used by currently successful commercial enterprises, and effectively exclude any support for projects likely to come up with technically or socially critical findings.

I would not want to exaggerate this influence. Like many other features of the post-modern social order, it tempts one to wax paranoid — especially if one knows, or believes one knows, some of the actual facts. My main point is that the order and influence is not unitary. It encompasses a plurality of strong social institutions, and here I include both state agencies and large commercial enterprises, each jostling the other for power.

But 'Science' is not one of those institutions. It is a loose mixture of research groups, expert communities, specialised bodies of knowledge, elaborate techniques, and instrumental facilities. Its potency is enormous. It enables the performance of great deeds by those who can pay its bills. But it has no organic strength or social clout as a whole.

Once upon a time, perhaps, 'Science' could be spoken for by the leading academic authorities. It was the president of the local national academy, one might say, who signed up to the 'contract with society'. This mythical deal is now coming up for revision. But who will negotiate it? As we have seen, the State is no longer in complete charge of Society, whilst the leading persona in the scientific world can no longer claim to represent a coherent, autonomous institution.

The production of scientific knowledge has been largely taken over by, or come under the influence of, the much stronger social formations that can afford to support it systematically. In some areas, as with the military-industrial-academic complex, the control is nearly absolute. Technoscience is simply the willing instrument through which the force of physical violence is monopolised and made available for exercise in society at large. In other areas, research is simply one of the means by which large firms compete for, and state agencies regulate, both economic and political power.

There remain only a few places in the post-modern social order where Science is still undertaken systematically as a distinctive activity. But even in those parts of academia that have not yet been commercialised, it survives very uneasily. Despite the grand scale of some of its instrumental facilities and global projects, it lives from hand to mouth. Its apparently expensive tastes are sustained by public and private patronage for which it must continually plead. It claims the right to do technical things its own way, but is in no position to offer any resistance to the commands of its patrons.

So the place of Science in contemporary society is not that of an independent institution. Half a century ago, at the end of the Second World War, it looked as if it was becoming one of the great 'Estates' of the Nation. Science — typified as academic science — seemed to stand alongside other 'Pillars of the Establishment' such as the Church, the Law, the Armed Forces, the Press and the City. The post-modern social order is somewhat differently structured. But in it the high official status of organised science, as such, has not been sustained. It does not occupy the seats of power. Cabinets nowadays don't always include a 'Minister for Science'. The corporate head of research doesn't always sit on the Board. The Admiral in command of weapon development isn't often a member of the Joint Chiefs of Staff.

Perhaps this was inevitable. As we have seen, the practice of research has become much more elaborately organised. But this development was far outmatched by the growth of the major structural organs of society. Even so-called Big Science is small beer for very big nations and international organisations managing vast military alliances, and for very big businesses trading globally. Relatively speaking, it costs them little to buy it up and bring it inside their tent.

Scientists complain that science has been 'industrialised',[36] 'commercialised',[37] 'militarised',[38] 'bureaucratised',[39] and 'commodified'.[40] What this really means is that it has been assimilated into the normal structures of industry, commerce, the military, the state bureaucracy, the market and other social institutions. But this is just another facet of the perfusion of science into our culture and its increasing incorporation in our regular social practices.[41] Furthermore, it signifies an increasing 'scientisation' of these pillars of society, from their lifeworld foundations right up to the top.

Is this development a matter for satisfaction? Should we not simply accept that the systematic production of scientific knowledge is a highly specialised and very risky activity, akin to foreign currency trading or the design of Formula 1 racing cars, which is best performed in the shelter of the mammoth organisations that are able to manage it, finance it and exploit its output to enhance their own powers? Or should we make some effort to retain a more autonomous scientific enterprise, capable of playing the non-instrumental roles that we have considered at length in earlier chapters? This will be the theme of the final chapter.

[36] Ravetz, J. R. (1971). *Scientific Knowledge and its Social Problems.* Oxford, Clarendon Press.

[37] Cozzens, S. E., P. Healey, et al., Eds. (1990). *The Research System in Transition.* Dordrecht, Kluwer.

[38] Salomon, J.-J. (2001). *Le scientifique et le guerrier.* Paris, Belin.

[39] Ziman, J. M. (1994). *Prometheus Bound: Science in a Dynamic Steady State.* Cambridge, Cambridge UP, Ziman, J. M. (1995). *Of One Mind: The Collectivization of Science.* Woodbury NY, AIP Press.

[40] Gibbons, M. and B. Wittrock, Eds. (1985). *Science as a Commodity: Threats to the open community of scholars.* London, Longman, Salomon, J. J. (1985). *Science as a commodity: Policy changes, issues and threats. Science as a Commodity: Threats to the open community of scholars.* M. Gibbons and B. Wittrock. London, Longman: 78-98, Salomon, J.-J. (2001). *Le scientifique et le guerrier.* Paris, Belin.

[41] Nowotny, H., P. Scott, et al. (2001). *Re-Thinking Science: Knowledge and the Public in an Age of Uncertainty.* Cambridge, Polity Press.

TWELVE

Science, Citizenship and Civil Society

The politics of knowledge

So we are back at politics again. The conventional discourse about the role of science in society does not go beyond arguing that it is a mighty performer in the *economy*. But in a competitive capitalist society, market success attracts disproportionate financial and managerial resources. These accumulate until they enable social control over people and institutions—i.e. *political* power. Governments, in their turn, require scientific capabilities to curb these forces. The affairs of the nation are settled in public and in private, conflicts where scientific knowledge is a potent weapon. Thus, organised science is not just an effective means of technological innovation and wealth production: as we have seen, it is now also a systemic element in the whole polity.

In an ideal capitalist world, this would be nothing to worry about. The plurality of companies competing for trade and influence would keep markets open and make knowledge freely accessible. Democratic institutions would hold the ropes, level the playing fields, maintain public facilities and protect weaker citizens from ill fortune. Science would just be a factor of production, like financial capital and labour, or a regulatory instrument like company law. Research would be undertaken by specialised enterprises, making a respectable living by selling their products and services to all and sundry.[1] The competitive plurality of these undertakings and their ways of knowing would ensure an abundant supply of reliable, disinterested knowledge.

Needless to say, the post-modern world is very far indeed from this ideal. Its global markets are dominated by multi-national oligopolies running their own private research and development facilities. Its political arenas are dominated by nation-states armed with the output of their closed technoscientific establishments. Governments employ further legions of scientists in support of their other respon-sibilities. These powerful social organisms exercise increas-ing influence over the independent research enterprises, the academic research groups, the state-financed research organisations that do most of the rest of the organised scien-tific work in our society. The means of knowledge produc-tion are now as highly concentrated in as few hands as the financial, industrial and political forces with which they are allied.

But doesn't the same apply to many other components of our culture? Think of the news media, literature, music, the arts, entertainments, leisure facilities, financial services, health services and consumer goods generally. In almost every case, a major fraction of the total supply comes from no more than half a dozen independent sources. I am not decrying what is also contributed by self-employed indi-viduals and what the economists call 'Small and Medium Enterprises'. It is to these, indeed, that our pluralistic culture owes much of its creative vitality. But they only con-tinue to exist by virtue of the economic, social and political rights won by centuries of popular struggle against earlier oligarchies.

To put it another way: the post-modern social order is remarkable for the material wealth, personal freedom, good health, educational standards, communication facilities, enjoyed by a large middle stratum of the population in eco-nomically developed countries. But it is also remarkable for

[1]   Kealey, T. (1996). *The Economic Laws of Scientific Research*. London, Macmillan.

the persistence of extreme inequalities of power, privilege and prosperity. These inequalities are observable in every scale, from the grotesque global range of *per capita* national incomes to the extremes of individual wealth and poverty juxtaposed in every large city. This pervasive inequity is a threat to both the moral legitimacy and the political stability of the whole system.

The same inequity applies to access to the power of organised scientific research. I am not referring here to the scientists themselves. However eminent they are, these are merely technical operatives, handing out knowledge as innocently as bank clerks hand out pound coins and dollar bills. I am referring to the bodies that fund, manage and eventually profit from these operations — the analogues of the banking corporations that determine in what enterprises and countries the big money is going to be invested.

I call this an inequity, because this is a 'knowledge society'. It is a culture where the welfare of individuals and of groups depends as much on access to knowledge as on access to money or to social status. Lack of it, for example, can prejudice the exercise of a citizen's rights under the law, or of his or her voice in democratic politics. As we have seen, science is now a major factor in everybody's lifeworld. People without some control over its sources feel like puppets on a string. To cure this malaise requires more than better public 'understanding' of science. By way of an analogy, think of the social reformers who argue for better popular 'understanding' of money. A more robust political tradition is to insist on a say in how it is actually produced and to whom it gets distributed. Exactly the same holds for scientific knowledge.

Most serious commentators on 'science and the public' are now saying much the same thing. This has evidently become a notable weakness in contemporary society. I would myself describe it as symptomatic of a much wider and deeper disorder in the whole system. But as I said at the beginning, I am not into general political theory, and I am not rooting for a radical alternative to the present socio-economic set-up. To my mind, utopian discourse on the ideal society is fruitless unless it indicates some vaguely plausible path from here to there. When it comes from natural scientists it also tends, as we have seen, to become much too technocratic. The ideal societies they design are often peculiarly nasty. They are professionally deformed to see the world in terms of 'problems' with correct 'solutions' which must be sought out and diligently imposed at any cost. The moral imperatives, paradoxes and uncertainties of all human existence are not on their radar screens.

That is one of the reasons why the *'republic of learning'* is an unsatisfactory model for the governance of society in general. This notion is attractive, especially to members of the scientific elite, but it misrepresents the logic of their situation. In reality, it applies solely to their public roles in the performance of academic science. As we have seen, this is a systematically stylised mode of knowledge production with remarkably reliable results. But its ethos is much too genteel to regiment the passions, prejudices, pretentions and personalities that interact so fiercely in our pluralistic culture. In any case, its normative hold over post-academic science is now 'more honoured in the breach than in the observance'.

Indeed, it must be admitted that most natural scientists, and quite a few social scientists, are not dissatisfied with post-modern society. They complain, of course, that it gives them too little money and too much paper work. But in their hearts they share its enthusiasm for managerial rationality, economic efficiency and technical efficacy. The great majority of them seem to fit easily enough into large-scale organisational frameworks, and get sufficient gratification out of making their expertise valuable to the state or the firm. Thinking about it makes us see that they almost invented globalism. The pluralism of the social order also harmonises with the pluralism of the scientific endeavour. Thus, the competitive funding of post-academic science matches the rivalrous spirit of research, and the winner-takes-all mentality of the free enterprise economy finds satisfaction in discovery races for Nobel Prizes. In a sense, science is better adapted to the social climate than ever before.

So it is no good expecting the scientific community to reform itself. Its characteristic response to political threats to its own autonomy — even to academic freedom in general — has been described as 'prudential acquiescence'.[2] Research scientists are not trained to think about how knowledge is used, and they have very little insight into their public image. If they enter politics at all, it is mainly through the policy and managerial interfaces. Sometimes, as we have seen, they become involved as experts in social and legal issues. Some are active in the big causes of our era — peace, the environment, poverty, and racism — but no more so than are other members of the intelligentsia.

As we have seen, the great majority of scientists and technical workers nowadays have little prospect of rising into the traditional elite. Indeed, they are not so much 'knowledge producers' as 'data technicians'.[3] Many of them put up with job uncertainty and poor pay for large parts of their careers. But their working conditions do not bring them into daily contact with the ills and injustices that breed general political discontent and dissidence. There was a period, between the two World Wars, when many leading scientists were very openly 'on the Left'. These were the founders of 'Scientific Humanism' in the uneasy days between the two world wars. But this surge of radicalism was evidently not systemic, for it has largely subsided. So scientists themselves are unlikely to initiate political action to loosen the grip of big business and the state on the organised production of scientific knowledge. As we shall see, the diverse constituents of 'civil society' also have need of this empowering resource: they are going to have to fight for it.

Science in an agonistic culture

Indeed, to 'fight' is the only word for it, albeit metaphorically. The public culture of post-modern society is *agonistic*.

[2]   Haberer, J. (1969). *Politics and the Community of Science*. New York NY, Van Nostrand.

[3]   Nowotny, H., P. Scott, et al. (2001). *Re-Thinking Science: Knowledge and the Public in an Age of Uncertainty*. Cambridge, Polity Press.

That is to say, it is first polemical and then combative. The Law protects diversity of opinion, multiplicity of property rights, and plurality of association. It forbids physical violence, but it protects *debate* — the right to strive with rivals in open argument. In ancient Greece, the standard setting for such debates was the city market place, the *agora*. But this is not just talking about the formal procedures of legislatures, law courts, and official enquiries. Nowadays many controversial issues are debated, and sometimes effectively decided, in a variety of unofficial arenas such as the media, conferences, lectures, company meetings and other public or semi-public assemblies.

The subjects for such dispute are endless. Some are incidental to commercial competition and technological innovation. These spawn questions about the perception, assessment and regulation of risk and thence into the domain of prophecy about hazards, and precautions against disaster. We argue interminably about how to deal with crime, ill health, poverty, illiteracy, homelessness, drug abuse, and a host of other social ills. Each of these ills generates an enormous burden of specific cases that have to be debated before they can be decided. Inequities of wealth, power and taxability still activate the political agenda. Ethnic, religious and other groups contend with one another for sectarian privileges.

In essence, ours is a culture devoted to *change*. Its watchwords are economic development, technological progress, political action, religious evangelism, and ideological revision. This dynamism inevitably breeds controversy. Some will gain from the changes, others may lose. But conflicts must be settled by the force of argument, not by the argument of force. That is the spirit of democracy and of the rule of law.

Thus, in every local arena of the multiplex market places (agora) of post-modern society, science is an effective weapon of attack or defence. The knowledge it produces is systematically organised, communally accepted, and already hardened by critical review. But as we have seen, it is not necessarily all-conquering. At base, it is founded on convincing *rhetoric* rather than compelling logic. Its power to persuade is limited by credible dissent and intrinsic uncertainty. Very often, what is reliably known to science is not relevant to the point at issue. Nevertheless, a well-founded scientific argument is an invaluable resource in orderly public dispute. Just think of the power of DNA profiling, for both the prosecution and the defence, in the criminal courts of the land.

Indeed, it is sometimes suggested that 'science' is so reliable, and so universally applicable, that it should be used to decide all legal and political issues. More modestly and plausibly, it is proposed that the relevant scientific facts and theories should be teased out and established by independent experts before any more general debate is opened. If this were done carefully, so it is claimed, then many instances of 'irrational' social discord could be fixed scientifically without the need for public controversy.

In practice, however, the concept of a 'Science Court' is of only limited utility. Should we be surprised? This is just another example of the technocratic fallacy. As we have seen, the world of research itself is characteristically agonistic. Consensus on broad principles and basic facts is not reflected in the specialised details. In any case, as we have seen, scientific knowledge is typically general, and requires highly questionable interpretation for 'local' application. So judging 'what is known to science' in any particular case is usually just as inconclusive as deciding the whole issue.

For this reason legal and political procedures are beginning to focus on the social mechanics of research. Lawyers and judges, politicians and civil servants, now require a clear understanding of how science works. The need to know what weight to give to its findings in particular cases. For example, courts of law and public tribunals are learning to distinguish between: (1) An established scientific 'truth'; (2) The near consensus of a research community; (3) A finding published in a peer-reviewed journal; (4) The considered opinion of a recognised expert; and (5) The say-so of just any professional 'scientist'. At one extreme we have knowledge that is effectively indisputable: at the other end of the spectrum, it usually merits less credence than ordinary common sense. People are coming to realise that the force of a scientific argument depends enormously on how much communal support there is behind it.

As we have seen, individual scientists often play a valuable social role as expert assessors, advisers or arbitrators on disputed issues. But even when they are just being consulted on details, they should *not* be considered infallible 'authorities'. Nor should they be thought to come to the place of decision already equipped with all the specialised knowledge required to reach some scientifically correct 'solution' to a complex 'problem'. It simply means that they are regarded as independent, morally responsible citizens who are well enough informed to grasp and weigh up the technical arguments on either side. Just as with judges and government officials, it is their honesty, impartiality and proven critical intelligence that wins public trust and legitimates their decisions. In a fast-moving culture, where delay is normally more costly than imperfect action, that level of integrity is often as good as it ever gets.

But a scientific argument is seldom invincible. Indeed, it usually just stimulates a horde of plausible counterarguments. Social conflicts similarly become polarised. Despite the pluralism of post-modern society, many-sided controversies are typically framed as public duels. It's the Prosecution against the Defence, the Government against the Opposition, and the Capital against Labour and Them against Us. So the scientific arguments are gathered together into two portfolios, labelled 'The Case For', and 'The Case Against'. The champions on the two sides become *advocates*, presenting their own arguments as logically unassailable whilst exposing the inconsistencies, uncertainties, defective data, *non sequiturs,* and even the downright 'terminological inexactitudes' of their opponents.

Our legal and political practices are indeed *adversarial*. A public controversy is customarily staged as a *debate*, culminating in a yes-no decision. This procedure might seem to harmonise with the disputatious spirit of science. But scientific controversies are not normally conducted in that format.[4] 'Debatable' scientific questions are, of course, frequently discussed very openly, and advocated with much vigour and not a little heat. But although the discourse often becomes strongly polarised, it is seldom structured as a stark choice between yea or nay. Many different points of view are articulated and set against one another, but all that happens in the end is that some gain wider credibility and kudos whilst others lose it.

Scientific expert witnesses often seem like fish out of water. They do not know how to breathe the oxygen of an adversarial process. They are accustomed to the moral support of a trusting audience, and are bewildered by apparently wilful attacks on their motives. Their uncertainties are misrepresented as blind ignorance. Their professional experience is discounted and their considered assessments derided as mere opinion. It is seldom a successful performance — or a happy experience.

The real point, though, is that adversarial processes require adversarial knowledge, and thus generate adversarial science and adversarial research. At first that seems to accord with Karl Popper's famous dictum[5] that the function of research is to refute conjectures. But this apparently negative attitude is normally counterbalanced within science by the positive intention of eventually producing well-tested knowledge. That is why academic science can foster peace and reconciliation in the aftermath of advocacy and competition. The high value placed on what is already agreed and on prospects for an eventual consensus enables some academics to play peculiarly influential roles as arbitrators and mediators in social conflicts. Members of the Royal Society have quite often taken up this burden, showing that this relic from distant and more formal days of

---

[4]   Ziman, J. (2000). 'Are debatable scientific questions debatable?' *Social Epistemology* 14(2/3): 187-199.
[5]   Popper, K. R. (1935 (1959)). *The Logic of Scientific Discovery*. London, Hutchinson, , Popper, K. R. (1963 (1968)). *Conjectures and Refutation: The Growth of Scientific Knowledge*. New York NY, Harper Torchbooks.

science can still find occasional uses for its formalities in their careful, published reports.

But a research project that is designed solely to demolish the case of a commercial or political opponent lacks this creative purpose. Furthermore, its context is very narrow and limited, so its outcome is unlikely to be of general application, and would typically be kept secret unless it had to be brought into action against a feared eventuality. What is worse, from the start it is quite explicitly partisan. As we have seen, these requirements are quite contrary to the norms of academic science.

On the other hand, 'defensive' research is a normal feature of technoscience. Any large organisation involved in a legal or regulatory dispute is bound to set its scientists to work on investigations that might be expected to provide results supporting its case. Indeed, if it is planning to launch a major technological innovation, such as a 'block buster' drug, it will have been undertaking research and taking out patents as a precaution against all conceivable challenges. This may also include projects probing the weaknesses of competing products, in the cases when push has to come to shove in a public arena. Conversely, official regulatory bodies have to be equally well equipped scientifically to contest the technical claims of such powerful organisations.

In a sense this is all just the good clean fun of industrial capitalism. But it does markedly tilt the playing field in favour of the party that is strongest in technoscience — that is, the one with the deepest pockets. Not only can big business afford expensive lawyers: it can also deploy lots of research results in support of its activities. Big government — if it is so minded — can hold its own in such disputes. But poorer contestants, with their much more modest scientific capabilities, are seriously disadvantaged.

Take, for example, the long-running dispute over the health risks of genetically modified crops. Suppose that certain varieties of these are being marketed by a very large multinational firm — let us call them PQLR Seeds, Inc. Suppose that the safety of these is being publicly questioned by a well established voluntary society — 'Greedy for Greens' say. No law has been broken, and there is no case for a civil action. And yet the issue is hotly contested in the media and other open arenas, with highly profitable consumer acceptance as the ultimate reward.

Now see how heavily the cards are stacked in favour of PQLRS. They have been working for years to develop their particular product and have already found out a great deal about both its positive and negative qualities. Their opponents, on the other hand, start only from a general doubt about the supposed safety of all such foods and can only cite a few contrary examples from other contexts. Academic research has produced a lot of theoretical knowledge that might prove to be relevant, but nothing specifically about the products in question. To make its case, GfG needs an intensive (and expensive) research programme focussed precisely on the points at issue. But that would be far beyond its means. So it happens on a number of occasions that a socially significant scientific question doesn't get properly thrashed out, and PQLRS strengthens its market power over suppliers and consumers, with a monopoly in research on its own operations.

It is usually argued that this type of issue, apparently threatening public health, should be taken up by the government. But can this be done without apparently favouring one or the other side in the dispute? British experience with a major government-sponsored study of the ecological effects of certain types of GM crops was not very encouraging. Serious political questions begin to be asked when the state research system is clearly engaged in adversarial technoscience. Is the government acting for, or against, big business? Anyway, as we have seen, the 'NIBA' that might have done this research was privatised long ago, and its remnants now constitute one of the divisions of PQLRS. And, as we have seen, post-academic science is now so entangled in commercial interests that its supposedly objective, expert assessments are no longer trusted by the court of public opinion.

In essence, the role of science in our adversarial post-modern culture is often unavoidably adversarial. It cannot always present itself as a 'neutral' third-party, to be trusted to resolve a dispute with magisterial authority.[6] Its function, rather, is to make the facts of the matter so compelling that a negotiated agreement becomes acceptable. A well-posed public debate between expertly-informed advocates is quite an effective way of getting at these facts. Indeed, despite the dangers of rhetorical excess, a lot of good science is actually tested in just that spirit.[7]

But such contests are worse than useless unless the contestants are fairly matched. And this, as we have seen, is no longer just a matter of ensuring equal access to *existing* knowledge. It also requires equity in the capacity to produce relevant *new* knowledge. That depends on who sets the agenda of research and has control of its outcomes. This question has become central to the place of science in post-modern society.

Ethics and values

The integration of science into society markedly increases the traffic between the research world and everyday life. Furthermore, this flow is not just one way. Not only does science speak freely in social contexts: nowadays, 'the context speaks back'.[8] As we have seen, this reflexive process usually takes place through professional practitioners, such as physicians and marketing consultants. The laboratory scientists mostly get these messages at third hand, and rely on their corporate or institutional managers to translate them into research programmes.

Adversarial technoscience, however, requires detailed attention to contextual voices. Research plans, their performance, and the presentation of their findings are all very sensitive to what 'people' say — that is, people quite outside the circles and clubs of scientific and technical expertise.

[6]   Nowotny, H., P. Scott, et al. (2001). *Re-Thinking Science: Knowledge and the Public in an Age of Uncertainty.* Cambridge, Polity Press.

[7]   Ziman, J. M. (2000). *Real Science: What it is and what it means.* Cambridge, Cambridge UP.

[8]   Nowotny, H., P. Scott, et al. (2001). *Re-Thinking Science: Knowledge and the Public in an Age of Uncertainty.* Cambridge, Polity Press.

The controversy occasioning the research may be of wide public concern. Its outcome may affect the interests of many groups besides the official contestants. For instance, GfG may not have many paid-up members, but surveys show that more than 10% of the population are sympathetic to its cause. And the fortunes of PQLRS will be watched anxiously by its employees, shareholders and suppliers. The course of the dispute will not go unnoticed by government ministers worried about industrial innovation and national competitiveness. Lawyers, politicians, theologians, media pundits and social activists of every persuasion will all express very positive views on what should be done.

Not, of course, that this babble of voices will contribute much to the scientific debate. In fact, they will mostly be quite ignorant, or grievously misinformed, about the epi-transformative suppressor gene and the apoptotic protolytic 2-5 enzyme (I've made that one up too) that might be so carcinogenic. They probably do have a vague notion of how such crops are cultivated, harvested and processed. But they hold strongly to *values* that are just as intrinsic to the controversy as the scientific facts. Life-world contexts always include the *moral*. They are regulated by *ethical* principles as well as by the laws of nature.

We do not need to catalogue the variety of beliefs, ideologies, predilections and opinions that come under this heading of 'the moral'. The simple fact is that this is a much more familiar component of commonplace knowledge than anything scientific. We all get to know about them as we grow up. In time we become aware of the complexity of the moral landscape through which we shall have to make our way. This landscape is not mapped out scientifically. So we learn to take our science with a pinch of ethical salt — that is, to adapt its truths to the imperatives of our other personal or communal values. In other words we, probably quite rightly assume that the findings of research are essentially *amoral* and not at all reliable in the 'ought' dimension. Indeed, to treat them otherwise would be to submit to a correspondingly amoral technocratic authority, whose algorithm is as likely to be devilish as sublime.[9]

But as we have seen, scientific knowledge is valued for its 'consensibility'[10] of the community of scientists working one any particular problem. It seeks to be acceptable as 'the truth' within as wide a community as is likely to take an interest in it, and have come to a consensus. To the trained scientific eye, it may even happen that all other values are out of focus. The obvious inconsistencies between widely held moral precepts are often confusing. Diligent search for a foundation for ethics still yields only incoherent fragments. In our pluralistic culture, contrary religions and ideologies are given equal protection by law. Even on a more secular level, perfectly sane individuals differ sincerely about how people should behave to one another and what really makes life good. On the focal plane of technoscience, all such considerations appear to be hopelessly blurred.

Nevertheless, somehow or another, adversarial research has to cope with these 'subjective' factors. In truth, the same applies to the study of any situation involving 'people'. That is why engineering and medicine are not just technosciences: they have developed elaborate professional practices to separate the designers and clinicians from the backroom boffins. But the active post-modern spirit will not tolerate such boundaries. We now realise that the local context, in all its buzzing confusion, cannot be represented by a few simple static conditions on the equations of motion. It interpenetrates and is a dynamical element of the system.

In principle, the human sciences — sociology and moral philosophy — ought to be able to tell us how to take account of the moral dimensions of social action and reaction. So the obvious first move is to build up research teams including experts from these human disciplines and/or to focus research programmes looking on these aspects of the situa-

[9]   Midgley, M. (1981). *Heart and Mind: The Varieties of Moral Experience.* London, Methuen, Midgley, M. (1992). *Science as Salvation.* London, Routledge.
[10]   Ziman, J. M. (1978). *Reliable Knowledge: An Exploration of the Grounds for Belief in Science.* Cambridge, Cambridge UP.

tion. That is the spirit in which the US Human Genome Project set aside a few per cent of its funds for academic research on 'ELSI' — the Ethical, Legal and Social Issues raised by the project as a whole.

In practice, however, this is not nearly enough. As we have seen, the human sciences themselves are often technocratic. In their striving to emulate the natural sciences, they too downplay the 'subjectivity' that makes people human. They too construct simplistic models that exclude precisely the moral sensibilities that activate ethical action. They over-egg scientistic concepts derived loosely from ambitious theoretical paradigms such as 'Evolutionary Psychology', 'Behaviourism', 'Sociobiology', 'Sociological Relativism', 'Economic Individualism' and so on. Indeed, the better they understand 'local' situations, the less direct use do they find for their generalised scientific knowledge.[11]

In fact, this is just one aspect of an endemic problem. The human sciences have a particular interest in agonistic situations. Public controversies take the place of experiments. The way that disputes are played out is a good test of their models and provides material for further theorising. But to understand what is going on, the researcher has to be able to empathise with the actors, and feel the moral forces that drive them. Can this be done without becoming identified as partisan?

Or, to put it the other way round: do you not repudiate your own moral principles when you are not actively supporting the side with whom you have come to sympathise? Isn't *participation* more truthful than 'objectivity'? Should the human scientist be expected to invert the professional role of the lawyer, and profess non-partisan views that she or he does not, at heart, believe?

This may well be less of a problem in practice than it is in principle. But there is a sense in which all serious social

[11] Lindblom, C. E. and D. K. Cohen (1979). *Usable Knowledge: Social Science and Social Problem Solving*. New Haven CT, Yale UP, Lindblom, C. E. (1990). *Inquiry and Change: The Troubled Attempt to Understand and Shape Society*. New Haven & London, Yale University Press.

research is implicitly adversarial. As we have seen, one of the major functions of non-instrumental research is to reveal unsuspected dangers. Indeed, as I shall argue later, our open, pluralistic culture is dynamically stabilised by the *critical* role of the human sciences. That is why defenders of the *status quo* tend to be suspicious of sociology. Those who persistently ask questions about the social order are surely very close to advocating change?

Nevertheless, the participation of human scientists in adversarial technoscience cannot cope fully with its moral dimensions. It is not enough for researchers to go round asking the contestants about their personal ideals, ethical norms, political aspirations, religious principles, ethnic commitments and scientific understanding. These 'values' are normally tacit, and may be quite misleading when articulated to order. The actors in a social drama may have perfectly clear ideas of what they are each trying to achieve in the particular circumstances, but very little insight into their own, or each other's, more general motives and preferences. The only way that these can be given adequate weight in the research is to involve the contestants themselves in its formulation and performance.

In effect, I believe that there are times when 'lay' persons — technically unqualified individuals — ought to be given an active, responsible part in the production of scientific knowledge. It is the context that speaks through them. They need to be included, along with the scientific experts, in the groups that draft and review research programmes and project proposals. As we shall see, they are often the key members of 'ethics committees' monitoring research performance. And they are in a position to decode the technical findings and locate them in the moral landscape of the life-world.

I am not talking here about public fora, citizen juries, advisory panels, focus groups and other mechanisms for tapping, articulating or arousing public opinion on contentious issues. Nor am I referring to the quasi-scientific attitudes and procedures that proliferate in our culture. As we have seen, post-modern 'folk science' is not necessarily 'pseudoscience'. Indeed, although 'common sense' now owes a lot to 'official' science it is still often much wiser in its own way.

No, what I have in mind is organised, expert, 'scientific' research projects, performed systematically at considerable expense, by skilled professionals using the most up-to-date intellectual and technical resources, to produce codified knowledge. In adversarial contexts, such projects are normally planned and carried out by multidisciplinary teams. What these teams should now have is a few 'lay' members. They should include responsible individuals whose principal qualification is that they are technically unqualified — whose 'discipline' is almost fulfilled by not having a discipline at all.

The task of these 'non-experts' is not just to articulate the partisan interests motivating the research, or to make sure — (perish the thought) — that it comes with the 'right' answers. True enough, they need to be personally familiar with and committed to the cause But their real role is to give the research process meaning in life-world terms. It is to testify to the tacit values that drive the dispute and to convey the norms that rule the actions of the contestants. It is to speak out, right inside the research team, for the moral context to which scientific expertise is usually so insensitive. They tend to call themselves 'professional ethicists'.

Research as an ethical pursuit

In the end, however, scientists working on value-laden problems in morally-saturated social environments cannot remain ethically 'neutral'. At the very least, they must attend sympathetically to the jabber of voices from the context to give meaning, purpose and ethical understanding to their research. And in extreme cases, no honest person could do the job at all without some level of personal commitment. As we have seen, this is well understood by experienced practitioners, and is becoming part of their professional training.[12] Laboratory-mode researchers would much prefer to stay quietly in their back rooms, but they too are being dragged out into the harsh light of social reality.

As I have argued in detail in another place,[13] this much greater sensitivity to ethical issues is one of the features of post-academic science. On the one hand, researchers working in academic institutions can no longer insist that their work is entirely non-instrumental, and therefore they need not be responsible for how it is used or misused. On the other hand, technoscientists cannot really go on claiming that they were only acting on the orders of their employers. So 'Scientific Ethics' has become a major topic in the discourse about 'Science and Society'.

Indeed, we have already encountered this theme in a number of places. I have continually presented science as essentially a moral enterprise. Thus, interpersonal *trust* is absolutely indispensable. I believe this is realised by most of the general public. They are disconcerted by situations where the *integrity* of science is questioned by 'conflict of interest' concerns including notions of bias, and where scientific *transparency* is negated in the context where research data are withheld. In the case of what appears to be proprietorial censorship this will nullify its perceived *sincerity*, and cases of plagiarism and fraud will certainly degrade its *honesty* in the eyes of concerned members of the public. Similarly the *authenticity* of its findings can be marred by commercial hype, especially if the *collegiality* of its communities is disrupted by bitter personal disputes. If there are cases where its *benevolence* is put in doubt by antisocial projects, and where its *independence* is threatened by zealous bureaucratic management it is bound to have some effect. The

---

[12] Stebbins, J. M. (2003). *Ethics and Social Responsibility in Engineering and Technology – Building Ethics into Professionalism*, New Orleans LA, Science and Engineering Ethics, 10, 201-432, (2004).

[13] Rotblat, J., Ed. (1982). *Scientists, the Arms Race and Disarmament.* London, Taylor & Francis, Ziman, J. M. (1998). 'Why must scientists become more ethically sensitive than they used to be?' *Science* 282: 1813-4.

members of the public may want to believe in science but are dismayed by episodes that seem quite out of keeping with its reputed virtues.

I am not suggesting that science is without vices, nor on the other hand that it has ever lived up to its noble reputation. Nor am I imputing that it is declining into dishonour. I am just saying that its traditional code of conduct is its greatest asset and the basis of its social influence. This is not just a code of professional *etiquette*: it is a potent *ethical* defence system. But as we have seen, it is largely tacit. There is no need for us to review once again the whole network of customary practices in which it is embodied. Only now, as the whole institution is being transformed by political and economic forces, are these being formally codified. But that may not be a good omen. An ethical norm that has to be defined legally is surely losing its moral strength.

Another conventional topic is the ethics of *research*. As science becomes more involved with society, its working procedures impinge more directly on people's lives. Biomedical research, in particular, cannot do without experiments on willing human subjects, not to mention unwitting animals, and especially non-human primates who are so close to us in their reflections on what is going on. How can their welfare be protected from undue interference in the name of progress and truth? Our post-modern society is peculiarly sensitive to such matters. So 'ethical committees', managerial bureaucracies, and expert professions have emerged to formulate and administer detailed guidelines and official regulations governing such situations.

Needless to say, this is one of the main arenas of social conflict associated with science. It is a context where 'values' often take precedence over 'technique'. The new discipline of *bioethics* has to take account of mundane factors such as the legal interpretation of voluntary consent or the psychiatric consequences of pain. But these have to be located in a framework of transcendental principles, such as the goodness of life, the evils of abortion, and the freedom to choose how to relate them. And these principles, whether or not they are grounded in religious doctrines, have not, so far, been amenable to clear consensus-making. This discipline is only 'scientific' in the general sense that it is an organised body of knowledge produced systematically by an academic community, but the other voices of consciousness are never quietened completely.

Nevertheless it is sometimes argued that all research scientists, along with practising physicians, engineers, lawyers and social workers, now need formal education in ethics. They must all be taught the rules and regulations governing their professional social roles, and should be given ample opportunities to rehearse a few of the moral dilemmas they will surely encounter in performing them. But these are not essentially different from any of the other 'local difficulties' in a conflict-riven world.

Ethical issues are not like well-posed scientific problems. They cannot be resolved 'rationally', or judged to be 'politically correct', by the application of a few grand principles. Nor can they be settled by knowledge of the 'great arguments' of the 'doctors and saints' who have disputed such matters for thousands of years. In practice, researchers, technoscientists and technical practitioners are like everybody else. Most of them are respectable citizens of well-ordered but irrevocably pluralistic societies. But as they grew up they each acquired a slightly different bundle of personal 'values'. They have learnt a little from experience about how to behave decently and they cannot really be taught formally to perform much better. If our scientific experts need systematic instruction in Ethics, then so do we all.

It is absolutely essential, however, that research scientists should not get the impression that the pursuit of 'scientific truths' exempts them from all other moral responsibilities. It is sometimes supposed — even in the human sciences — that research is conducted in another room, whence the natural and social worlds are observed through a one-way mirror. On the contrary, as we have seen, research is very often highly purposeful. As science comes into closer contact with society, it becomes a mode of social or humanistic *action*. The *intention* with which it is performed is at least as morally significant as the mode of performance.

As we have seen, post-academic science is entangled in a web of economic and political interests and institutions. These have societal agendas which shape the research projects they decide to foster and fund. A researcher who undertakes such a project cannot deny commitment to its explicit goals. Of course she may argue that she is merely a technician, a hired hand. But that would rob her of her principal asset as a scientist — her standing as an independent and morally responsible person.

Scientists, alas, are too often 'in denial', as some people put it, about this aspect of their vocation. Funding bodies do not advertise their more questionable purposes, and scientists seeking patronage have little cause to question them and consider this to be dangerous to their own purposes. Nevertheless, as we have seen, the exposure of unacknowledged 'conflicts of interest' can be devastating for professional researchers as it also is for members of the public with tender consciousness. Matters become far worse if any of the goals of the research are found to be antisocial — that is, against the interests of society in general.

The 'strong programme' in scientific ethics makes these goals public, and faces scientists and the angry members of the public with the moral implications of working to achieve them. In many cases, this is not such a difficult task. No right-minded person nowadays can seriously approve research designed to encourage people to smoke more tobacco, or to help speed maniacs avoid prosecution. Although these are not illegal, they are obviously extremely damaging to the public good. But one soon becomes embroiled in much more open moral debates. Is research on performance-enhancing drugs, for example, essentially unethical, or does this apply only to their illicit use in competitive sports?

This programme inevitably leads back into the general ethical agora of our culture. As we have seen, it introduces ideologies and doctrines that cannot be covered by scientific paradigms. It acknowledges contradictions and dilemmas that *cannot* be resolved by scientific methods. Of course there is a vast body of discourse on the subject, ranging from deep scholarly analysis to high spiritual insights. Nevertheless, in this domain — that is, life as we live it — we are all wanderers and pilgrims, guided by inconsistent moral maps or by unreliable intuitions towards ill-defined personal goals.

There are no generally accepted criteria by which a particular project can be judged to be completely 'ethical' or 'unethical'. It just has to be accepted that debates about 'science and ethics' cannot be closed. All that can reasonably be asked of any scientist who is possibly entering into antisocial research is to become fully informed of the moral issues that it raises, and to choose conscientiously between any of them if and when they seem to be in conflict.

What is truly irresponsible is to close ones eyes to all such issues. Take, for example, the case of military research. Like all good people, scientists do everything they can to mitigate tensions and conflicts in society. Sometimes they have unusual opportunities or special skills for doing so. But their peaceful aspirations are frustrated by the close connections of science with war. As we have seen, a large fraction of technoscience — and therewith, much of post-academic science — is devoted to the invention, development, design and production of military hardware. This is clearly an activity that has serious ethical implications for the whole scientific enterprise. Yet the fact that these are scarcely ever discussed in the scientific community shows that the scientists are still far from achieving consensus. This is not a call for more work which must lead to agreement between the scientists. It may be painful to find that ethical problems cannot be resolved to everyone's satisfaction, but that conclusion in itself is a salutary achievement.

Of course this is a topic on which views range widely. Some hold such employment to be quite unacceptable. Many agree with the Nobel Prize winner Joseph Rotblat that it is incompatible with the peace-loving principles of scientific endeavour. Many others hold that it is only justifiable in particular circumstances, such as a direct threat to the security of ones fellow citizens. A wider group find it acceptable to work for the armed forces of their own countries but would refuse to serve the international armaments trade. And the great majority of scientists are deeply ashamed that some of their colleagues are actively adding extra dimensions of horror to the most terrible events of our era.

The general ethical arguments for and against each of these positions are well known and not at all abstruse. They may not be of great concern to the general public, who are already somewhat sceptical about the high-mindedness of science. But why are they so seldom debated amongst scientists? It is true that they raise questions to which there are no universally valid answers. I certainly don't have a definite opinion on how to resolve the dilemmas they frame. But by not even discussing these dilemmas the scientific community seems to be 'in denial' of a major stain on its carefully nurtured moral reputation.

One reason for this silence may be that military technoscience is almost entirely in the hands of very large enterprises. As we have seen, it is performed by the agencies of national governments and by major industrial corporations. Furthermore, this 'military-industrial-scientific complex' is global. This makes it difficult to escape entirely from its influence. Any natural scientist or engineer in its employ is contractually bound to serve its interests. But these interests are also felt throughout the scientific world. A large proportion of the research work in post-academic science is directly or indirectly indebted to the defence sector of society—if only in small ways—for ideas and material support.

We are back once more at the problem of ensuring rhetorical equity in a public controversy. Individuals and small groups who might be minded to challenge the morality of this form of research are overawed by the massed political and economic ranks of their opponents. As we shall see, some institutionalised source of countervailing power is needed to fill this deficit in our post-modern culture.

Meanwhile, however, the ethical case against military technoscience is not being pressed publicly. Unfortunately, this leaves the impression that the place of science in society is indeed to operate as a willing partner of the military and of industry. If that were true, it would not only be deplorable on ethical grounds. It would also put our whole democratic, pluralistic social order at risk. Not only would it mean that the vital non-instrumental functions of science could no longer be performed. It would imply that the state itself, including the whole apparatus of representative government, was under the thumb of this unholy trio.

That radical inference may surely be exaggerated. But it is the public duty of the eminent persons who speak for science in order to rebut it convincingly. And that does not seem feasible without a serious discussion within the scientific world of its own ethical dimensions. Until now, this discussion has tended to be far too uncritically celebratory: in this post-modern era of weapons of mass destruction it needs far more to be totally realistic and sober.

Civil society countervails

The discordant values and ethical norms that are disputed in the post-modern market place are primarily personal. They are what we each desire, whether as affluent consumers, loving parents, responsible citizens or compassionate and spiritual neighbours. But these preferences are all so fragmented, contradictory and mixed up that they cannot be satisfied by individual action. People have to come together, agree on what they want, and concert their efforts to achieve it. That is what political parties are for. Even the much-vaunted commercial entrepreneur has to 'conspire' with many stake-holders to bring his goods to the market-place.

The essence of a pluralistic society is not just that it permits *individual* freedom: it is that it could also allow its members to combine to achieve their *collective* ends. The monstrosity of totalitarianism, whether socialist or capitalist, theocratic or technocratic, is that a huge body people, inside or outside the governing elite, usurps this right and systematically denies it to any other social formation. The tragic history of modern times has taught us that a thriving *civil society* is the best indicator of general political well-being.

Unfortunately, this is a buzz word that means too many different things. Originally the adjective *civil* referred to 'what pertains to the citizen in his *ordinary* capacity' as distinct from 'various words expressing specific departments'.[14] Thus, it was originally a term for those features of the social order that were *not* military, political, criminal, or ecclesiastical.

But a civil society is not just an amorphous crowd of 'civilians'. Its elementary units are not just 'people' but groups of people which together makes up the state. It is 'joined up' and purposeful in detail, for it seeks to make a Civil Society, however small that group may be. Some of its constituents, like nation-wide political parties and popular charities, comprise a significant fraction of the population; others, like minor religious sects and local sports clubs, are very small. Some have corporate charters, large bank balances and numerous employees; others rely on the voluntary labour and financial contributions of their members and supporters. It is, you might say, a 'dusty term' for what it stands for. Above all it is pacific, looking for peace both for the state as a whole, and for the individuals that comprise it. In between those limits it will contain organisations campaigning internationally for human rights and devoted to the good of humanity; others, such as trade unions and neighbourhood associations, are looking more closely to their own interests.

Some would have it that this relationship should be *moral*. They attribute to civil society an *emancipatory* spirit.[15] They note, for example, the way that many non-governmental organisations — NGOs — have burst out of their national

---

[14] *Oxford English Dictionary.*
[15] Keane, J. (2003). *Global Civil Society?* Cambridge, Cambridge University Press.

boundaries and linked globally into a powerful humanistic network of publicly acclaimed bodies with considerable influence on international affairs. Organisations such as Amnesty International, the Red Cross, the Green Crescent, Oxfam, Médécins sans Frontières and world-wide groups of TV celebrities, working bravely against the political grain to bring to the people a sense of justice, sustenance and freedom. Others strive for sustainability and ecological diversity, or pit themselves directly against global capitalism, which they hold to be a menace threatening human welfare. In effect 'Civil Society' is seen to be in the vanguard on every item of the progressive agenda. That is not an easy role to play.

Ernest Gellner, in his small but potent book on *Conditions of Liberty: Civil Society and its Rivals* (1996) produces some simple and valuable ideas which are both constructual — advice for builders of new societies — and also advice for keeping such a society on the rails once it is in motion. Gellner writes ' *the simplest, immediate and intuitively obvious definition, which also has a good deal of merit, is that Civil Society is that set of diverse non-governmental institutions which is strong enough to counterbalance the state'*. In this way it makes it possible to live with the accoutrements of the state. As we have seen, many people hold that science has a similar mission. It too is a global enterprise devoted to the advancement of truth and knowledge. It is celebrated as the mainspring of progress, and the prime source of improvements to the human condition. But that is not an easy path to follow. There are plenty of institutions which will fight for their own interests, but where would the arbitrator be found to protect the Civil Society once it is formed?

The civil society needs to combat ancient cruelties with critical rationality. It would need to set an example of orderly, creative controversy and also international cooperation for common goals. So it ought to be the natural ally of the state, both locally and globally, but only *if it can fulfil its role of preventing other local powers from 'preventing the Civil Society from dominating and atomising the rest of society'.*

It may be that neither of the partners in this coalition — the Civil State or Post-academic Science — is as coherent and high minded as we might have hoped. Of course I am not saying that 'science' is innately antisocial. But as we have seen, it has not always been a unitary undertaking driven by a transcendental ethos. Its metaphysical claims need to be vigorously questioned. Its technocratic 'solutions' are often considered more inhumane than the 'problems' they are tackling. (The ongoing warfare in Iraq is certainly pointing out to us all the miseries that preventing other local powers from *atomising the society* may bring.) Some institutional components of the civil society are quite disconnected from reality. Others are mere tools of the government and /or multinational corporations whose activities are so greatly deplored. Although scientific 'progress' can be extraordinarily beneficial, it requires continued and responsible ethical guidance if it is to be 'progressive' in the best political sense.

At the same time, although the existence of a civil society is the essential leaven of democratic pluralism, it is not a single-minded institution with a moral mission. It is not wholly *'agin* the government', nor entirely 'anti-capitalist', nor completely 'peace-loving', nor uniformly 'green'. The interests and goals of its innumerable and diverse aims may seem hopelessly contradictory and confounded. It is riven with religious, cultural, ideological and other sectarian conflicts. Even well run, high-minded NGOs with very similar missions may find it difficult to cooperate with them.

Civil society then, is not so much a 'movement' as a widely dispersed forum. It is the post-modern equivalent of the town meeting. But instead of giving each citizen a separate voice, it collects like-minded opinions and proclaims them energetically. It is thus a dynamic source of innovation, benefaction, criticism, protest, ethical norms and other societal forces. Above all, it empowers individuals and amplifies their assaults on the political and economic monoliths of our society. Through allowing countervailing domination by the state and by other entrenched institutions, it binds together and reaffirms the democratic social order.[16]

The constituents of civil society, small and large, big-hearted or small-minded, traditional or radical, are thus continually embroiled in controversy. In an ideal world, they would only be fighting for justice and equity against the other powers-that-be: but in practice much of the conflict is against one another. That, as we know all too well, in this disrupted first century of the new millennium, is the stuff of pluralistic politics.

That is where science may come in again. As we have seen, many of the issues over which people dispute are about the applications of technoscience. It is not always an effective instrument of advocacy. Access to relevant research findings may sometimes be the determining factor in a public debate and the ability to undertake appropriate research is a powerful adversarial resource.

In other words, science is now a vital adjunct to civil society. It not only facilitates philanthropy and arouses action. It is an essential part of the armoury of every would-be public combatant against the civil society. But as we have seen, all too often it is governments and commercial corporations that command its battle tanks and its smart missiles. A barrage of potent knowledge claims belches forth from their elaborate instruments, professional research teams, and electronic databases. How can a voluntary association of like-minded people ever prevail against them?

The organs of civil society are not quite powerless. In a genuinely open society they have a constitutional right of access to 'information', much of which may be scientific. More significantly, a great deal of scientific knowledge is continually being produced and published in the open literature of the web. Indeed, as we have seen, science — especially academic science — has traditionally operated as an *independent* source of images, attitudes, critiques, and warnings. In practice, this vital non-instrumental role is

[16] Gellner, E. (1994). *Conditions of Liberty. Civil Society and its Rivals.* London, Penguin.

largely performed through, and also to the benefit of, the civil society.

In essence, post-modern political, economic and cultural pluralism requires both widespread devolution of power and widespread dissemination of knowledge. Civil society is the means by which these functions are combined, performed and come to reinforce one another. Campaigning organisations with particular missions — for example, protecting wildlife or curbing the devastation of AIDS — can use the internet to obtain the latest research findings from books and journals and make them known, when appropriately simplified, to their supporters. They also use these findings in policy debates and legal disputes. It is the role of science, as one of the guardians of *public knowledge*, to enable such people to resist the bulldozers of state and of corporate technoscience.

To a tidy mind this does not seem a very satisfactory way of achieving 'public understanding of science'. In fact, professional experts are often appalled at the way that half-baked research claims which only just pass peer review are sometimes publicised and presented as established scientific truths by partisan groups. But civil society also includes relatively impartial organisations such as learned societies whose mission it is to try to 'put the record straight'. At times, of course, they too need to be controverted, although not — hopefully — to the same degree as commercial advertising or government propaganda.

The main thing is that questionable assertions are indeed questioned, alternative theories are given a hearing, and official authorities challenged. That is how emancipatory social goals are fought for and achieved, and also how scientific knowledge is tested and established. The agonistic politics of civil society mirrors, and links with, the agonistic procedures of research communities. This is a 'space in which science meets and interacts with many [...] agents, where institutions overlap and interact, and where interests, values and actual decisions to be taken are being discussed, negotiated, fought over and somehow settled'.[17] Here indeed is a dynamic, close-coupled interface between science and society.

In the modern 'information society', knowledge is power. But as we have seen, in the post-modern 'knowledge society', the power to produce knowledge is paramount. The diverse bodies that make up civil society can be given legal access to established scientific results. But they have puny research resources by comparison with their corporate and state opponents. They seriously lack, and desperately need, the means to acquire reasonably reliable, scientifically validated information on a great variety of highly technical matters. They ought not to have to rely on whatever happens to emerge out of the research system. They need to be able to initiate research projects relevant to their political missions, and have full access to their findings.

This type of knowledge can only come from people and institutions that are reasonably independent of corporate and state control. The public have unconsciously become accustomed to relying on academic science to support the civil society in this way. Unfortunately, as we have seen, post-academic science doesn't do non-instrumental research in this spirit. Much of its work is still very helpful to the systematically heterogeneous set of non-governmental organizations, voluntary associations, religious groups, not-for-profit companies and charitable foundations, that counterbalance big government and big business. But because these bodies have no direct influence over the agenda of research, they are seriously limited in the use to which its results can be put.

It is not enough to talk vaguely about greater popular 'participation' in science, or making scientists more 'ethically sensitive' or 'socially responsible'. Civil society badly needs its own serious research capacity. Good science is a sophisticated professional activity. And it requires managerial frameworks and substantial investments. It is

---

[17] Nowotny, H., P. Scott, et al. (2001). *Re-Thinking Science: Knowledge and the Public in an Age of Uncertainty*. Cambridge, Polity Press.

embedded in elaborate social structures. It is the work of *organisations* and not of isolated individuals. The immense human resources of civil society may always get fragmented and discordant. Very few of its constituents can afford to run their own research facilities or to pay the full market price for hiring them. They have to rely on what is done for them by fairly well resourced, and more firmly based bodies.

Civil society is not only one of the major 'sectors' of the post-modern social order. It is also one of the most important 'trading zones' between science and society. Of course it depends on the research activities of various independent, sympathetic, knowledge-producing institutions. As we have seen, the advent of post-academic science represents a radical structural transformation of the whole research system. The future place of science in society will thus be determined by whether such institutions continue to be fostered or at least be allowed to survive. This applies, in particular to their traditional location in 'academia'. The spotlight of our study thus turns to the university and the various social formations associated with it.

Academia as an independent estate

We have already ventured into academia at a number of points. But now we need to see it as one of the pillars of post-modern society. The deep historical roots, traditional practices and intangible operations of the university make it seem insubstantial. In reality, it is a solid, orderly social institution, with a variety of distinctive functions. Remember that a substantial and increasing fraction of the population spend several years of their lives under its wing.

Many of the social roles of science are actually performed in, or by, the university. Historically speaking, systematic scientific research is a late addition to its other intellectual responsibilities. Here is where knowledge is not just 'produced': it is also conserved, catalogued, reproduced, refined, and eventually disseminated throughout society. These interwoven functions give the whole institution great stability and permanence. Scientific progress will render the most advanced academic curriculum and the most brilliant discovery obsolete within a few decades. But the libraries, laboratories and lecture halls will continue to receive their annual cohorts of students, served by a braided network of attested scholars and hard-pressed administrators.

Let us deconstruct that a little. The 'academic sector' is not just one great bureaucratically ordered system, rationally subdivided into geographical campuses and subject faculties. As we have seen, it is an untidy market place of strenuously competing individuals and ideas.[18] This intangible agora is stabilised by the condensation of its activities into separate *universities* — the colleges, institutes and academies where researchers work together to share facilities, and dispute for the attention of audiences.

Over the centuries these institutions have variously proclaimed that they are dedicated to the furtherance of religion, moral education, the training of the higher ranks of society, the advancement of knowledge, the dictatorship of the proletariat, technological innovation, national competitiveness. But what they show in the long run is their ability to survive through 'interesting times'. This is only partly due to the educational services that continue to be needed by successive political regimes. It also reflects the community spirit of their members. For all their notorious internal dissensions and cumbersome out-of-date ceremonials, universities are amongst the most permanent social formations of historical times.

Nevertheless, the academic sector remains intensely competitive. Universities vie with one another for students, resources, faculty and sporting trophies. And apart from football stars, the trump cards in this game are eminent scholars and scientists. So the non-academic members of the university board of governors quiz their academic colleagues:

---

[18]   Ziman, J. (2002). The microeconomics of academic science. *Science Bought and Sold: Essays in the Economics of Science*. P. Mirowski and E.-M. Sent. Chicago, University of Chicago Press: 318-40.

Q.   *What makes a scientist eminent?*

A.   Outstanding discoveries.

Q.   *How do they make discoveries?*

A.   With good research facilities.

Q.   *What would tempt this particular eminent scientist to move here?*

A.   Much better research facilities.

Academic science is the driving force of the research university. As we have seen, it celebrates individual achievement, as recognised by communal peers. That sounds admirable in principle. But by its emphasis on originality it fosters extreme specialisation. (The quiz session now continues)

Q.   *So what did this lady actually discover?*

A.   It's difficult to explain, but a world expert that I happened to meet recently in Korea tells me it might eventually revolutionise the whole subject.

Q.   *Should we teach it to students?*

A.   Oh, no, not for some years, I guess: it's much too advanced.

Q.   *What was so special about it, then?*

A.   Well, it was a really superb piece of science, if you see what I mean.

In effect, the post-modern university does not have a direct corporate interest in the substance or significance of the knowledge that is produced on its premises by its academic employees. All it cares about is that this work should be well regarded by those technically competent to assess it. A Nobel Prize is a welcome adornment to the brand image, but not the sign of a marvellous addition to our scientific 'world picture'. What a university truly appreciates is the money that then flows in to fund yet more research. If it has large financial resources, these can be profitably invested in the laboratory buildings, experimental apparatus, technical staff and other facilities required to attract some more good researchers. Of course university financial officers haggle with grant-giving bodies over the 'indirect' costs of research projects. But they know very well that the welfare of their whole enterprise depends much more on its reputation as a source of new knowledge than as a place where the established wisdom is treasured and transmitted.

It is easy to grow sentimental—or cynical—about the relationship between academic institutions and the knowledge they process. What I am getting at is that they are *politically* very weak. They do not operate corporately as major stakeholders for science in society. For all their brave talk about academic freedom and autonomy, they almost never get into open conflict with the state or with big business. They are not even notably successful at defending their own ethos. There are notorious cases, for example, where a great university has failed to protect individual faculty members from persecution for their socially unwelcome views.

Indeed, the notion of academia as an autonomous 'estate' of the polity does not stand up to close inspection. Their students and alumni belong to the respectable, responsible strata of society. In many fields of knowledge they are closely involved with technoscientific and managerial practice. Even when they are not officially organs of the state, universities are typically dependent financially on government and corporate patronage. As we have seen, post-academic science enhances this dependence.

There are now very few academic institutions in the world with sufficient economic and social capital to 'go it alone'. But then, which way would they go? Their experience as self-governing scholarly communities does not provide them with a collective stance in relation to society at large. In any case, this stance would most likely be conservative rather than radical. As we have seen 'prudential acquiescence' is typical of their response, individually and institutionally, to more purposeful societal forces.[19]

---

[19] Haberer, J. (1969). *Politics and the Community of Science*. New York NY, Van Nostrand.

Academic pluralism and social dissent

And yet the academic ethos is not an empty ideology. For more than a century, a great many scientists have been able and willing to practise its precepts. As we have seen, it still shapes many of the features of post-academic science. It is true that universities never adhered to it systematically in the past, and they may be even less bound by it nowadays. Many institutions of higher education, for example, have always been closely connected with technical and business practice, and perform directly instrumental research as a normal service to their corporate or governmental patrons. But much of the knowledge emanating from even the most utilitarian 'institutes of technology' is still being produced in the 'academic' mode.

This is because academic science is not just a 'tradition'. It emerged, and still functions, in a potent cultural form. Metaphorically speaking, it is an assembly of interlocking markets.[20] It is a global network of multiply-connected communities with no general system of governance, weak at every particular point *but extremely resilient overall*. Even though most of the big players in this game — universities, funding agencies, commercial firms, etc. — would like to organise their research differently, they must follow its unwritten rules. And that means that the basic principles of the academic ethos are still very much alive in the minds and hearts of most university researchers.

For example, universities compete for any high-achieving faculty. The measure of achievement is communal esteem for published work. So 'publish or perish' is the recipe for success — just as the academic ethos requires. Again, since research findings will not get published unless they are reasonably original and critically robust, these norms are installed in the institutional culture.

This ethos is not prescribed from the top. It is not inscribed in tablets of stone. But it is continually enacted

---

[20]   Ziman, J. (2002). The microeconomics of academic science. *Science Bought and Sold: Essays in the Economics of Science*. P. Mirowski and E.-M. Sent. Chicago, University of Chicago Press: 318-40.

locally in the procedures and scholarly principles that are customary in each discipline. A university department that is known not to respect these principles will fail to attract competent staff. Soon it will soon lose its reputation as a place where 'good science' is done.

Competition in the academic marketplace[21] thus sets standards for working conditions, as well as for the people selected to do the work. In particular, the global rivalry between research universities reinforces the norm of 'disinterestedness'. To win an appointment in another institution — perhaps in a foreign country — a scientist *must not* seem bound by local interests. Why should anyone be given a post in Australia if all their research activities continued to be associated with a project located in Scotland? Well, university folk often do that sort of thing, but only on condition that they can demonstrate the generality of their perspective. What holds for the one who is in Perth (Australia) must seem to be equally relevant to the other one who is in Perth (Scotland).

Thus, to keep in the swim, academic scientists must endeavour to appear independent of church and state, commerce and industry, political party and ethnic community. Call this philosophical 'objectivity' if you like, but notice that it not only attaches to ideas and attitudes; it also colours the roles, tasks, and social structures at the research coalface. The general public are suspicious of the 'ivory tower' — elitism typically fostered in academia. As we have seen, academic science is very often dangerously detached from the realities of the life-world. It downplays local, contextual knowledge in favour of universal generalities. But academic institutions have good reason to protect their most 'creative' researchers and research groups from interests and concerns that might harm their standing in the wider scholarly world.

As we have seen, however, this world is irredeemably pluralistic. Not only is it divided and sub-divided into innu-

---

[21]   Caplow, T. and R. J. McGee (1958). *The Academic Marketplace*. New York NY, Basic Books.

merable disciplines and research specialities. Within each little domain of knowledge, and its paradigms, their standard bearers contend and clash. Indeed, a research field where there is consensus is the one most likely to be moribund, or in the grip of an unhealthy intellectual fashion! So the scientific standing of an individual scholar or research team is not like the rating of a Judo champion or chess grandmaster. It is fluid, indeterminate, and very much a matter of opinion. When a prize is to be awarded or an appointment to be made, the 'world authorities' are consulted. But their assessments of achievement, let alone of 'promise', are very seldom unanimous. And you would have been wrong to believe them when they all agreed that this middle-aged pedant (e.g. Max Planck) would never make a worthwhile discovery, or that this field of research (e.g. high-temperature superconductivity) was now worked out. Neither prognosis proved to be correct — quite the contrary.

As a result, good universities are not entirely staffed with diligent conformists. They contain many hopeful heretics, persistent *provocateurs*, dreamy dissidents and stubborn sceptics. The problems on which they research are not all 'normal' in the sense of being defined within an established paradigm.[22] Some of the projects they undertake are highly speculative or wilfully revolutionary. Of course, much of this mould-breaking effort is fruitless, and the great majority of the would-be pioneers do not break through to celebrated discoveries. Indeed some of them, if the truth be known, turn out to be sad no-hopers who are not even competent teachers. To a ruthless sociological eye, these wasted lives are unavoidable by-products of the knowledge-creation process.

In other words, academic managers — senior professors, department chairs, faculty deans and university presidents, rectors and vice-chancellors — have to be enlightened pluralists. They know that it is their duty to tolerate and

[22]   Kuhn, T. S. (1962). *The Structure of Scientific Revolutions*. Chicago IL, U of Chicago Press.

support intellectual dissent, even when it seems implausible or unruly. Universities are emboldened to take a gamble on an eccentric scholar, an unconventional research project or even an *avant garde* institute or department. When this gamble succeeds, there is a long-term payoff in public reputation; when it fails, all that is usually lost is just the original financial stake. Even when they don't positively proclaim their attachment to academic freedom and the unfettered search for truth, it suits their image to be known as sources of original and/or heretical ideas, havens of critical technical dissent, and reservoirs of free-thinking expertise.

In most branches of the natural sciences and their associated technologies dissensus on the campus is not a matter of much concern to the society. The academic machinery of a well-founded discipline is perfectly capable of absorbing unconventional views, putting them through the mill of communal criticism, and eventually commending to the general public those that are deemed to be sound. The corporate, regulatory and market mechanisms of industrial technoscience also, are geared to select imaginative innovations, and discard those that are useless or dangerous.[23] As we have seen, society has eventually benefited from the exposure of hazards by 'out-of-area' researchers in fields such as ecology or climatology. And, in some sciences, such as cosmology, internal controversy may distress the participants (but gives positive pleasure to the spectators!)

In the human sciences, however, the situation is much more delicate. The topics that are debated in academia are often closely linked to political and social issues. Intellectual disciplines such as history, sociology, economics, social psychology, social anthropology and moral philosophy are not merely highly disputatious. They also engage with the values for which people are prepared, sometimes, to fight and die. As we have seen, these values are central to any scientific understanding of what is going on, and inevitably they become subjects of academic controversy. In the end,

---

[23] Ziman, J. M., Ed. (2000). *Technological Innovation as an Evolutionary Process*. Cambridge, Cambridge UP.

the public agora extends into the lecture room, and the professors become partisans of reaction or reform. Naturally enough, they often emerge on the political scene as supporters of civil society in campaigns for social change.

To a truly enlightened pluralist, this is not a problem. As we have seen, one of the non-instrumental roles of academic science is to offer informed criticism of the social order. Through their research activities, universities have traditionally provided the public with independent scientific expertise for the democratic control of technocratic power. But they have to survive in relatively obscure societies. So they often get into trouble for sheltering social, moral, or political free-thinkers. In reality, these attitudes are shared by only a small proportion of their faculty members. Nevertheless in these areas of knowledge they have to be extremely cautious in public about any institutional support they may seem to give to the unconventional causes advocated by their more deviant academic employees.

Unfortunately, the transition to post-academic science has widened the scope of this cautious attitude to dissent. Critical investigations and unorthodox views in the natural sciences also come under public scrutiny. Corporate patrons do not welcome sceptical opinions on commercially sensitive research findings. Ethical campaign groups denounce scientific projects that seem to transgress their moral principles. Government agencies do not fund research programmes that might question their policies. The dissenters are not dismissed: they are just quietly starved of the resources they would need to establish their critique.

Science for the citizen

This book has been about establishing a fruitful and harmonious relationship between the citizens of a pluralistic polity and their science. We can now see that this depends on strengthening the connections between academic science and civil society. Here are two components of our culture that seem to be made for each other. Their common stance is critical, creative and pluralistic, rather than conformist, conventional and monopolistic. They operate openly and legally, both by persuasion rather than by force. They are both loose-linked networks without an overall framework of authority. In their different ways, they both seek the welfare of society, whilst remaining independent of its major centres of power.

Furthermore, they have complementary features. Academic science has the capacity to produce the rigorous, reliable knowledge that civil society needs to develop its policies, defend them and put them into effect. Civil society, on the other hand, can provide the values and passions that make that knowledge desirable and morally principled. Of course these characteristics are also uncomfortably contradictory. Highly disciplined knowledge, aware of its own limitations, can paralyse action. Incoherent life-world precepts, zealously applied, can blind scientific researchers to inescapable truths. But these contradictions are intelligible, and can be overcome by sincere debate.

It is not surprising, then, that there is an open interface between these two segments of our culture. As we have seen, civil society draws heavily on the open knowledge base of academic science, and often makes use of its professional expertise. At the same time, many academics are sympathetic to the societal goals of various civil society organisations, and they work positively for them. I am not just talking of the grandly radical opinions expressed by a few well known scholars and scientists. I am thinking of quiet research programmes that ask critical questions about government policies, or expose commercial exploitation and corporate corruption. The sober articles, lectures, interviews and books that open up and articulate these issues, are as valuable to civil society as its own, often more vociferous, efforts to produce social change.

For example, the 'Peace Movement' is a major sector of civil society in many countries round the world. Ever since Hiroshima, an immense variety of organisations, large and small, have been campaigning, agitating, lobbying and occasionally taking 'direct action', in opposition to governmental policies on national and international security, especially in relation to the use of nuclear weapons. Much of the information on which they base their arguments comes from the published work of academic political scientists, historians, technologists and engineers. Furthermore, several universities have established departments or institutes of 'Peace Studies' or 'War Studies' where research on these matters is conducted, and students taught about them. These are not very different in practice from the 'Think Tanks' where NGOs produce reports on contentious issues. Although the academics are careful not to appear overtly partisan, their personal role in providing knowledge, advice and strategic guidance to the relevant civil society organisations is never, and cannot be, concealed.

Nevertheless, the connections across the interface are informal, voluntary and *ad hoc*. They depend very much on the traditional rights of academics to speak what they know and act as free citizens in an open society. In many cases, they are reporting the findings of 'their own work', undertaken as much to establish their scholarly credentials as to develop a political or social theme. In other cases, it is research funded by independent foundations or by campaigning NGOs. If the going gets rough, the university can always disclaim any official responsibility for the views thus expressed.

The truth is that universities are not really autonomous. Scholarly and scientific communities are not free-standing collectives. The production of knowledge is not a self-sustaining social enterprise. Academic scientists are not completely at liberty to follow the dictates of their own consciences. Their licence to speak freely does come with the research facilities required to justify their dissent. Research institutes are not obliged to be socially responsible. In sum, academia is very far from being a segment or inseparable partner of civil society.

Furthermore, as we have seen throughout this work, the transition to post-academic science has made all such relationships much more difficult and questionable. Universities have always been beholden to patrons to support their research capabilities. Nowadays they are frankly in the market for more research funds. The salaries, equipment, buildings, facilities and administrative costs required to produce scientific knowledge mostly have to come from other bodies. Faculty members are even expected to tout for grants to support their own research projects. Here, as elsewhere in life, *those who pay the pipers call the tunes.* Academic science is directly dependent on these funding bodies and is shaped by their policies and preferences. It can no longer operate independently, along with other non-governmental and non-commercial organisations, in the pluralistic domain outside the state and corporate sectors of society.

Thus, at a time when science and society are being drawn into an ever-closer union, there is a deep flaw in their relationship. The spotlight of science does not get turned on for a variety of issues of deep concern to many people. The knowledge that ought to illuminate many social, ethical and political problems only comes in fragments from unfocused sources. Even the unplanned flashes of insight traditionally generated by non-instrumental academic science are being filtered out.

Democratic governments are not unaware of this knowledge gap. To their credit they often try, quite sincerely, to fill it. Their funding agencies initiate programmes of research on various areas of public concern or social need — crime, drugs, medical ethics, housing, education, food safety, etc. As we have seen, this research is of great societal value. But it cannot be disconnected from government policies and state interests. Even when these do not directly influence the findings, they frame the questions whose answers are being sought. For example, what government agency could afford politically to support an open-ended research project investigating in depth the military-industrial-scientific complex that has so much power in many countries?

Needless to say, commercial technoscience cannot perform this function. It may be a major actor in some of these issues, but is not to be trusted to resolve them. In any case, the market does not put a value on matters where it does not see a possibility of profit. These are questions of public good, not opportunities for private gain.

On the other hand, lavish spending on academic research is not a realistic solution. As we have seen, everybody would benefit if researchers working in academia were liberated from some of the managerial constraints of post-academic science. But they would not necessarily devote themselves to socially relevant projects. Problems promising the communal rewards of scientific discovery are much more attractive. The self-organising processes of the research world are not weighted towards the study of pressing public issues. The redefinition or reorientation of scientific disciplines to cover these issues is a laborious process, which is surprisingly difficult to drive by any acceptable political means.

The orthodox view is that this gap can be bridged by involving ordinary people more closely in the reception and social application of the knowledge produced by science. Focus groups, citizens juries and other talk shops reveal the reactions of typical members of the public to techno-scientific change. They thus articulate the complex of values surrounding technical and cultural innovation. In practice, the participants in these discussions are very often members of organised pressure groups. Civil society thus has its say in how science is introduced into society.

But by the time that scientific knowledge is ready for application, the research train has already arrived. It is too late to stop it, or direct it to another destination. As we have seen, the key to power in the post-modern, 'knowledge society' is the research agenda. The initiation, formulation and selection of research projects is now a major channel of political influence. What we get to know — and thereby can do — depends as much on what we seek to know as on what we eventually discover.

Here is where civil society ought to have a much stronger formal voice. In particular it should have the main responsibility for defining and representing the 'context of implementation' — the life-world setting of both language and ethics where the expected results of research are to be put into practice. This involves much more than consultation over the ethics of a proposed investigation. It means having a hand in the formulation of research questions and protocols. It implies power to define problems that ought to be looked into, and to initiate scientifically sound research on them. In essence, civil society should be as influential in this vital social role as is corporate commerce.

Itemised in detail this is a very tall order. Research systems would have to be substantially reformed. At the very least, it would affect the membership and charter of a great many committees. Novel voices would have to be listened to at all levels, from the governing councils of funding agencies and academic institutions down to the review panels in research specialties. It would also involve serious consideration of how to tap into the amorphous network of disparate organisations that constitute this 'civil society'. Mechanisms would also have to be installed to prevent very small but zealous groups, such as anti-vivisectionists, from routinely blocking any projects they regarded as unethical.

Nevertheless, the basic principle of this proposal is clear and simple. Although its objectives are radical, they could be implemented piecemeal. Unfortunately that means that it would have to present innumerable detailed targets for opposition. Furthermore, it would have to be developed country by country to accord with the local political culture. That process, like the contents of many sections in this book, would indeed be a tall order — but immensely well worth doing.

Bibliography

ABRC (1987). *A Strategy for the Science Base*. London, HMSO.

Albrow, M. (1996). *The Global Age: State and Society beyond Modernity*, Cambridge: Polity Press.

Arthur, W. B. (1994). *Increasing Returns and Path Dependency in the Economy*. Ann Arbor MI, U of Michigan Press.

Auyang, S. Y. (1998). *Foundations of Complex-System Theories: in Economics, Evolutionary Biology and Statistical Physics*, Cambridge: Cambridge University Press.

Barnes, B. & Edge, D., ed. (1982). *Science in Context: Readings in the sociology of science*. Milton Keynes: Open UP,

Barrow, J. D. (1991). *Theories of Everything: The quest for ultimate explanation*. Oxford, Oxford UP.

Basalla, G. (1988). *The Evolution of Technology*. Cambridge, CUP.

Beck, U. (1986/1992). *Risk Society: Towards a New Modernity*. London, Sage.

Bernal, J. D. (1939). *The Social Function of Science*. London: Routledge.

Bodmer, W. (1985). *The Public Understanding of Science*. London, The Royal Society.

Bok, D. (2003). *Universities in the Marketplace: The Commercialization of Higher Education*. Princeton NJ, Princeton University Press.

Bush, V. (1945). *Science – The Endless Frontier: A Report to the President on a Program for Postwar Scientific Research*. Washington DC, National Science Foundation.

Calogero, F. (2000). 'Might a laboratory experiment destroy planet earth?' *Interdisciplinary Science Reviews* 25(3): 191-202.

Campbell, D. T. (1960). 'Blind variation and selective retention in creative thought as in other knowledge processes.' *Psychological Review* 67: 380-400.

Campbell, D. T. (1979). 'A tribal model of the social system vehicle carrying scientific knowledge.' *Knowledge: Creation, Diffusion, Utilization* 1: 181-201.

Caplow, T. and R. J. McGee (1958). *The Academic Marketplace*. New York NY, Basic Books.

Castells, M. (2000). *The Rise of the Network Society*. Oxford, Blackwell.

Cohen, J. and I. Stewart (1994). *The Collapse of Chaos: Discovering Simplicity in a Complex World*. Harmondsworth, Penguin.

Collins, H. and T. Pinch (1993). *The Golem: what everyone should know about science.* Cambridge, Cambridge UP.

Collins, H. M., Ed. (1982). *Sociology of Scientific Knowledge: A Source Book.* Bath, Bath UP.

Constant, E. (2000). The evolution of war and technology. *Technological Innovation as an Evolutionary Process.* J. Ziman. Cambridge, Cambridge University Press: 281-98.

Cooper, W. (1952). *The Struggles of Albert Woods.* London, Jonathan Cape.

Cozzens, S. E., P. Healey, et al., Eds. (1990). *The Research System in Transition.* Dordrecht, Kluwer.

CSS (1986). *UK Military R&D: Report of a Working Party of the Council for Science and Society.* Oxford, Oxford University Press.

CSS (1989). *The Value of Useless Research.* London, Council for Science and Society.

Cziko, G. (1995). *Without Miracles: Universal Selection Theory and the Second Darwinian Revolution.* Cambridge MA, MIT Press.

Dasgupta, P. and P. A. David (1994). 'Towards a new economics of science.' *Research Policy* 23: 487-521.

Douglas, M. (1986). *Risk Acceptability According to the Social Sciences.* London, Routledge & Kegan Paul.

Ezrahi, Y. (1980). Utopian and pragmatic rationalism: the political context of scientific advice. *Minerva* 18, 110-31,

Ezrahi, Y. (1990). *The Descent of Icarus: Science and the Transformation of Contemporary Democracy.* Cambridge MA: Harvard UP.

Forman, P. (1971). 'Weimar culture, causality and quantum theory, 1918-23.' *Historical Studies in the Physical Sciences* 3: 1-116.

Frascati Manual (2002): *Proposed Standard Practice for Surveys on Research and Experimental Development.* Organisation for Economic Co-operation and Development (OECD).

Galison, P. (1997). *Image and Logic: A Material culture of Microphysics.* Chicago IL, University of Chicago Press.

Garfinkel, H. (1967). *Studies in Ethnomethodology.* Englewood Cliffs NJ, Prentice-Hall.

Geertz, C. (1973). *The Interpretation of Cultures.* New York NY, Basic Books.

Gellner, E. (1994). *Conditions of Liberty. Civil Society and its Rivals.* London, Penguin.

Gibbons, M. and B. Wittrock, Eds. (1985). *Science as a Commodity: Threats to the open community of scholars.* London, Longman.

Gibbons, M., Limoges, C., Nowotny, H., Schwartzmann, S., Scott, P. & Trow, M. (1994). *The New Production of Knowledge.* London: Sage.

Greenberg, D. S. (1969). *The Politics of American Science.* Harmondsworth, Penguin.

Grinnell, F. (1992). *The Scientific Attitude.* New York & London: Guilford Press.

Gross, P. & Levitt, N. (1994). *Higher Superstition: The Academic Left and its Quarrels with Science.* Baltimore MD: Johns Hopkins UP,

Gross, P. M., Levitt, N. & Lewis, M. W., ed. (1996). *The Flight from Reason*. New York NY: New York Academy of Sciences.

Haberer, J. (1969). *Politics and the Community of Science*. New York NY, Van Nostrand.

Hagendijk, R. P. (2004). 'The Public Understanding of Science and Public Participation in Regulated Worlds.' *Minerva* XLII(1): 41-59.

Hagstrom, W. O. (1965). *The Scientific Community*. New York NY, Basic Books.

Haldane, J. B. S. (1927). *Possible Worlds and Other Essays*.

Holton, G. (1993). *Science and Anti-Science*. Cambridge MA: Harvard UP.

Horgan, J. (1996). *The End of Science: Facing the limits of knowledge in the twilight of the scientific age*. New York, NY: Addison Wesley.

Irvine, J. and B. R. Martin (1984). *Foresight in Science*. London, Francis Pinter.

Jagtenberg, T. (1983). *The Social Construction of Science*. Dordrecht, Reidel.

Jasanoff, S., G. E. Markle, et al., Eds. (1995). *Handbook of Science and Technology Studies*. Thousand Oaks CA, Sage.

Kant, I. *Prolegomena to any Future Metaphysics*. This reference is from: Horgan, J. *The End of Science: Facing the limits of knowledge in the twilight of the scientific age*, New York, NY: Addison Wesley, 1996., who got it from Rescher, N., *Scientific Progress*, Pittsburgh PA: University of Pittsburg Press 1978.

Kauffman, S.A. (1993). *The Origins of Order: Self-Organization and Selection in Evolution*. Oxford, Oxford University Press.

Kauffman, S.A. (1995). *At Home in the Universe: The Search for Laws of Complexity*. London, Viking Press.

Kealey, T. (1996). *The Economic Laws of Scientific Research*. London, Macmillan.

Keane, J. (2003). *Global Civil Society?* Cambridge, Cambridge University Press.

Knorr-Cetina, K. D. (1981). *The Manufacture of Knowledge: An Essay on the Constructivist and Contextual Nature of Science*. Oxford, Pergamon.

Kuhn, T. S. (1962). *The Structure of Scientific Revolutions*. Chicago IL, U of Chicago Press.

Langer, Susan (1957). *Philosophy in a New Key*. Cambridge MA, Harvard UP.

Latour, B. and S. Woolgar (1979). *Laboratory Life: The Social Construction of Scientific Facts*. London, Sage.

Lifton, R. J. (1993). *The Protean Self: Human Resilience in an Age of Fragmentation*. Chicago IL, University of Chicago Press.

Lindblom, C. E. (1990). *Inquiry and Change: The Troubled Attempt to Understand and Shape Society*. New Haven & London, Yale University Press.

Lindblom, C. E. and D. K. Cohen (1979). *Usable Knowledge: Social Science and Social Problem Solving*. New Haven CT, Yale UP.

Lovelock, J. (1991). *Gaia: The Practical Science of Planetary Medicine.* London, Gaia Books.

Lovelock, J. E. (1979). *Gaia: A new look at life on the Earth.* Oxford, Oxford UP.

Mansfield, E. (1991). The social rate of return from academic research. *Research Policy.*

McLuhan, M. (1962). *The Gutenberg Galaxy.* London, Routledge, Kegan Paul.

McNeill, W. H. (1982). *The Pursuit of Power.* Chicago IL, University of Chicago Press.

Merton, R. K. (1942 [1973]). The Normative Structure of Science. *The Sociology of Science: Theoretical and Empirical Investigations.* N. W. Storer. Chicago IL, U of Chicago Press: 267-78.

Midgley, M. (1981). *Heart and Mind: The Varieties of Moral Experience.* London, Methuen.

Midgley, M. (1992). *Science as Salvation.* London, Routledge.

Midgley, M. (2001). *Gaia: The next big idea.* London: Demos.

Mitchell, G. R. (2000). *Strategic Deception: Rhetoric, Science and Politics in Missile Defence Advocacy.* East Lancing, MI, Michigan State University Press.

Mokyr, J. (1990). *The Lever of Riches.* New York NY, Oxford UP.

Nelson, R. R. (2003). The advance of technology and the scientific commons. *Phil. Trans. R. Soc. Lond.* A 361: 1691-1708.

Nowotny, H., Scott, P. & Gibbons, M. (2001). *Re-Thinking Science: Knowledge and the Public in an Age of Uncertainty.* Cambridge: Polity Press.

Olson, M. V. (2002). The Human Genome Project: A Player's Perspective. *Journal of Molecular Biology* 319(4): 931-42.

Oreskes, N. (1999). *The Rejection of Continental Drift: Theory and Method in American Earth Science.* Oxford, Oxford University Press.

Pacey, A. (1992). *The Maze of Ingenuity: Ideas and Idealism in the Development of Technology.* Cambridge MA, MIT Press.

Pickering, A. (1984). *Constructing Quarks: A Sociological History of Particle Physics.* Edinburgh, Edinburgh UP.

Pickering, A. (1995). *The Mangle of Practice: Time, Agency and Science.* Chicago IL, U of Chicago Press.

Pickstone, J. (2000). *Ways of Knowing: A New History of Science, Technology and Medicine.* Manchester, Manchester University Press.

Polanyi, M. (1962). The republic of science: its political and economic theory. *Minerva* 1(1): 54-73.

Popper, K. (1945). *The Open Society and its Enemies.* London: Routledge and Kegan Paul.

Popper, K. R. (1935/1959). *The Logic of Scientific Discovery.* London, Hutchinson.

Popper, K. R. (1963/1968). *Conjectures and Refutation: The Growth of Scientific Knowledge.* New York NY, Harper Torchbooks.

Price, D. J. d. S. (1963/1986). *Little Science, Big Science — and Beyond.* New York NY, Columbia UP.

Primavesi, A. (2000). *Sacred Gaia: Holistic theology and earth system science*. London, Routledge.

Ravetz, J. R. (1971). *Scientific Knowledge and its Social Problems*. Oxford, Clarendon Press.

Ravetz, J. R. (1977). *The Acceptability of Risks*. London, Council for Science and Society.

Roberts, R. R. (1989). *Serendipity: Accidental Discoveries in Science*. New York NY, Wiley.

Rose, S. (1997). *Lifelines: Biology, Freedom, Determinism*. London, Penguin.

Rotblat, J., Ed. (1982). *Scientists, the Arms Race and Disarmament*. London, Taylor & Francis.

Salomon, J. J. (1985). Science as a commodity: Policy changes, issues and threats. *Science as a Commodity: Threats to the open community of scholars*. M. Gibbons and B. Wittrock. London, Longman: 78-98.

Salomon, J.-J. (1999). *Survivre à la science: Une certaine idée du futur*. Paris, Albin Michel.

Salomon, J.-J. (2001). *Le scientifique et le guerrier*. Paris, Belin.

Schütz, A. and T. Luckmann (1974). *The Structures of the Life-World*. London, Heinemann.

Shinn, T. (2002). The Triple Helix and New Production of Knowledge: Prepackaged Thinking on Science and Technology. *Social Studies of Science* 32(4): 599-614.

Simon, H. (1996). *The Sciences of the Artificial*. Cambridge MA, MIT Press.

Snow, C. P. (1964). *The Two Cultures: and a second look*. Cambridge, Cambridge University Press.

Solomon, J. (1992). *Getting to Know about Energy in School and in Society*. London, Falmer Press.

Stebbins, J. M. (2003). Ethics and Social Responsibility in Engineering and Technology — Building Ethics into Professionalism, New Orleans LA, *Science and Engineering Ethics*, 10, 201-432, (2004).

Toulmin, S. (1972). *Human Understanding, Vol. 1*. Oxford, Oxford University Press.

Turner, S. P. (1990). Forms of patronage. *Theories of Science in Society*. ed. S. E. Cozzens and T. F. Gieryn. Bloomington IN, Indiana UP: 185-211.

Weber, M. (1918/1948). Science as a vocation. From *Max Weber*. H. H. Gerth and C. W. Mills. London, Routledge & Kegan Paul: 129-56.

Weinberg, A. M. (1962). Criteria for scientific choice. *Minerva* 1(2): 158-71.

Wheeler, M., J. Ziman, et al., Eds. (2002). *The Evolution of Cultural Entities*. London

Wolpert, L. (1992). *The Unnatural Nature of Science*. London, Faber & Faber.

Wynne, B. (1991). Knowledges in context. *Science, Technology & Human Values* 16, 111-121.

Ziman, J. (1968). *Public Knowledge: The Social Dimension of Science*. Cambridge, Cambridge UP.

Ziman, J. (1976). *The Force of Knowledge*. Cambridge, Cambridge UP.

Ziman, J. (1978). Bounded science: the prospect of a steady state. *Minerva* 16: 327-39.

Ziman, J. (1978). *Reliable Knowledge: An Exploration of the Grounds for Belief in Science.* Cambridge, Cambridge UP.

Ziman, J. (1979). *Deciding about Energy Policy.* London, Council for Science and Society.

Ziman, J. (1980). *Teaching and Learning about Science and Society.* Cambridge, Cambridge UP.

Ziman, J. (1981). What are the options? Social determinants of personal research plans. *Minerva* 19: 1-42.

Ziman, J. (1981). *Puzzles, Problems and Enigmas: Occasional Pieces on the Human Aspects of Science.* Cambridge, Cambridge UP.

Ziman, J. (1983). The collectivisation of science. *Proc. Roy. Soc.* B 219: 1-19.

Ziman, J. (1984). *An Introduction to Science Studies.* Cambridge, Cambridge UP.

Ziman, J. (1987). *Knowing Everything about Nothing: Specialization and Change in Scientific Careers.* Cambridge, Cambridge UP.

Ziman, J. (1991). Academic science as a system of markets. *Higher Education Quarterly* 45: 41-61.

Ziman, J. (1994). *Prometheus Bound: Science in a Dynamic Steady State.* Cambridge, Cambridge UP.

Ziman, J. (1995). *Of One Mind: The Collectivization of Science.* Woodbury NY, AIP Press.

Ziman, J. (1996). Is science losing its objectivity? *Nature* 382: 751-4.

Ziman, J. (1996). Postacademic science: Constructing knowledge with networks and norms. *Science Studies* 9: 67-80.

Ziman, J. (1998). Why must scientists become more ethically sensitive than they used to be? *Science* 282: 1813-4.

Ziman, J. (2000). Are debatable scientific questions debatable? *Social Epistemology* 14(2/3): 187-199.

Ziman, J. (2000). *Real Science: What it is and what it means.* Cambridge, CUP.

Ziman, J. (2002). No Man is an Island. *Hermeneutic Philosophy of Science, Van Gogh's Eyes, and God: Essays in Honour of Patrick A. Heelan, S.J.* B. E. Babich. Dordrecht, Kluwer: 203-18.

Ziman, J. (2002). The microeconomics of academic science. *Science Bought and Sold: Essays in the Economics of Science.* P. Mirowski and E.-M. Sent. Chicago, University of Chicago Press: 318-40.

Ziman, J. (2003). Emerging out of nature into history: the plurality of the sciences. *Phil. Trans. R. Soc. Lond.* A 361: 1617-33.

Ziman, J. (2003). The Economics of Scientific Knowledge: Review of Shi (2001). *Interdisciplinary Science Reviews* 28(2): 150-2.

Ziman, J. and M. Midgley (2001). Pluralism in science: a statement. *Interdisciplinary Science Reviews* 26(3): 153.

Ziman, J. Ed. (2000). *Technological Innovation as an Evolutionary Process.* Cambridge, Cambridge UP.

Ziman, J., Sieghart, P. & Humphrey, J. (1986). *The World of Science and the Rule of Law.* Oxford: Oxford UP.

Index

problems, defining 84
problem-solving expertise 85, 86, 114
progress, concept of 114
  moving boundary of 58
project planning 164-5
  projects and lack of opportunities for spontaneous development 166
protoplasm 218, 221
pseudoscience 308
psychiatry 4
public accountability 53
public goods 74-5, 267, 272-3
public knowledge 53, 89, 320
public science 111
public: and science 2, 6, 7, 234
  attitudes towards science and scientists 2, 19, 174, 183, 215, 231, 310, 314, 327
  interaction with science 307-8, 334
  misunderstanding of science 177
  popular thought and scientific ideas 238
  public criticism of science 203-4
  public participation in scientific knowledge 307-8, 321
  understanding of science 149, 252-4, 294,
publications, scientific and technoscientific 68, 125, 287
  and authenticity 150, 200
  and social norms 191
  see also peer review
'Publish or Perish' 95, 326

Qualified Scientists and Engineers (QSEs) 253, 279-80
Quality Adjusted Life Years 64
Quantum Theory 110, 195, 221
quarks 101, 194, 202, 221

R&D, corporate and industrial 75, 80-1, 118, 130, 202, 281, 286, 288
  aspects of R&D defined 24
  managerial culture and 120-1
  and 'Mode 2' knowledge 82
  necessary infrastructure for 87
  government and 121

RDD&D (Research, Development, Design and Demonstration) 4, 265, 282
real world and scientific theories 202
reality and non-resemblance to the laboratory 208-9
received wisdom 63
Red Cross 317
reductionism 203, 223-5
reductionist fantasies 221
regulatory bodies, official 301
regulatory science 184-5
relativists, sociological 216
relativity 221
religion 39
republic of learning 295
Republic of Science 20
research
  agonistic nature of world of research 298
  amorality of research findings 304
  applied 24, 25
  bandwagon of 115
  basic 24
  'being scientific' 64
  categories of 24-5
  changing research cultures 115, 133
  and commissions of inquiry 284
  and competition for power 289-90
  consumer research 207
  contemporary 233
  in context 211-13
  contract research 163
  and credibility 110
  defensive research 301
  and development, aspects of 24ff
  ethics of research 310, 334
  experimental 24
  as a factor of development 9-10
  function of research 300
  instrumental function 24-6, 60
  modern companies and 283-4
  multinational research projects 277
  non-instrumental and technoscience 86ff
  network 287